Polymer Science and Technology

www.novapublishers.com

Polymer Science and Technology

Luxury Animal Fibres Part 1: Hair Fibres from Goats
Riza Atav, PhD and Lawrance Hunter, PhD
2023. ISBN: 979-8-88697-998-5 (Softcover)
2023. ISBN: 979-8-89113-046-3 (eBook)

Nanocomposite Hydrogels and Their Emerging Applications
Mohammad Sirousazar, PhD (Editor)
2023. ISBN: 979-8-88697-675-5 (Hardcover)
2023. ISBN: 979-8-88697-895-7 (eBook)

Far Infrared and Terahertz Spectroscopy of Polymers
Valery A. Ryzhov
2022. ISBN: 978-1-68507-662-7 (Softcover)
2022. ISBN: 978-1-68507-830-0 (eBook)

A Study of Polymer Dynamics by Solid-State NMR
John Nobleman
2021. ISBN: 978-1-53619-715-0 (Hardcover)
2021. ISBN: 978-1-53619-803-4 (eBook)

What to Know about Lignin
Jalel Labidi, PhD, M. Özgür Seydibeyoğlu, PhD,
and María González Alriols, PhD (Editors)
2021. ISBN: 978-1-53619-152-3 (Hardcover)
2021. ISBN: 978-1-53619-222-3 (eBook)

More information about this series can be found at
https://novapublishers.com/product-category/series/polymer-science-and-technology-series/

Rıza Atav
and Lawrance Hunter

Luxury Animal Fibres

Part 1: Hair Fibres from Goats

www.novapublishers.com

Copyright © 2023 by Nova Science Publishers, Inc.

DOI: https://doi.org/10.52305/GLRK7183

All rights reserved. No part of this book may be reproduced, stored in a retrieval system or transmitted in any form or by any means: electronic, electrostatic, magnetic, tape, mechanical photocopying, recording or otherwise without the written permission of the Publisher.

We have partnered with Copyright Clearance Center to make it easy for you to obtain permissions to reuse content from this publication. Please visit copyright.com and search by Title, ISBN, or ISSN.

For further questions about using the service on copyright.com, please contact:

                                       Copyright Clearance Center
Phone: +1-(978) 750-8400   Fax: +1-(978) 750-4470   E-mail: info@copyright.com

NOTICE TO THE READER

The Publisher has taken reasonable care in the preparation of this book but makes no expressed or implied warranty of any kind and assumes no responsibility for any errors or omissions. No liability is assumed for incidental or consequential damages in connection with or arising out of information contained in this book. The Publisher shall not be liable for any special, consequential, or exemplary damages resulting, in whole or in part, from the readers' use of, or reliance upon, this material. Any parts of this book based on government reports are so indicated and copyright is claimed for those parts to the extent applicable to compilations of such works.

Independent verification should be sought for any data, advice or recommendations contained in this book. In addition, no responsibility is assumed by the Publisher for any injury and/or damage to persons or property arising from any methods, products, instructions, ideas or otherwise contained in this publication.

This publication is designed to provide accurate and authoritative information with regards to the subject matter covered herein. It is sold with the clear understanding that the Publisher is not engaged in rendering legal or any other professional services. If legal or any other expert assistance is required, the services of a competent person should be sought. FROM A DECLARATION OF PARTICIPANTS JOINTLY ADOPTED BY A COMMITTEE OF THE AMERICAN BAR ASSOCIATION AND A COMMITTEE OF PUBLISHERS.

Library of Congress Cataloging-in-Publication Data

ISBN: 979-8-88697-998-5

Published by Nova Science Publishers, Inc. † New York

Contents

List of Figures vii
List of Tables xi
Preface xiii
Acknowledgments xix

Chapter 1   **An Overview of Luxury Animal Fibres** 1
                1.1. Hair Origin   4
                1.2. Secretion Origin   12

Chapter 2   **Luxury Fibres Obtained from the Goats** 19
                2.1. Mohair Fibres   20
                      2.1.1. Historical Background of Mohair 22
                      2.1.2. World Production of Mohair Fibres 26
                      2.1.3. Harvesting Mohair Fibres
                                and Factors Affecting Yield 35
                      2.1.4. Classification of Mohair Fibres 37
                      2.1.5. Microscopic Properties of Mohair Fibres 43
                      2.1.6. Physical Properties of Mohair Fibres 48
                      2.1.7. Chemical Properties of Mohair Fibres 65
                      2.1.8. End-Uses of Mohair Fibres 68
                2.2. Cashmere Fibres   77
                      2.2.1. Historical Background of Cashmere 79
                      2.2.2. World Production of Cashmere Fibres 81
                      2.2.3. Harvesting Cashmere Fibres and Factors
                                Affecting Yield 87
                      2.2.4. Classification of Cashmere Fibres 92
                      2.2.5. Microscopic Properties of Cashmere Fibres. 93
                      2.2.6. Physical Properties of Cashmere Fibres 96
                      2.2.7. Chemical Properties of Cashmere Fibres ... 105
                      2.2.8. End-Uses of Cashmere Fibres 107
                2.3. Fibres of Goat Hybrids   111
                      2.3.1. Cashgora Fibres 111

References 119

Index 133

About the Authors 137

List of Figures

Figure 1	Main sources of luxury animal fibres	6
Figure 2	Fineness (diameter) ranges of various luxury animal fibres	9
Figure 3	Main sources of secreted luxury fibres	13
Figure 4	Angora (Mohair) goat	21
Figure 5	Distribution of angora production by country	27
Figure 6	Share (%) of global mohair production according to country as at 2021	30
Figure 7	Geographical distribution of Angora goats in Turkey in 2010	31
Figure 8	Fluctuations in South African mohair prices between 1970 and 2000	35
Figure 9	Mohair fibres; (a) cross section, (b) longitudinal view (fine and coarse fibre) and (c) view of the scale layer (surface structure) (X8000)	44
Figure 10	Mohair fibre structure	44
Figure 11	Types of medullae; (a) unbroken lattice (wide) (b) simple unbroken (c) interrupted (d) fragmented	47
Figure 12	(a) Cross-sectional (b) longitudinal (c) the scale layer (surface structure) of heavily medullated (i.e., kemp) fibres	48

List of Figures

Figure 13	The effect of goat age and associated body weight on mohair yield and fibre properties	51
Figure 14	Price changes according to the fineness of South African mohair in 1999	52
Figure 15	Price change according to the length of South African mohair in 1999	54
Figure 16	Example of white and coloured mohair fibres	63
Figure 17	Coloured Angora goats	63
Figure 18	(a) IMA Mohair Mark and (b) South African Mohair Mark	71
Figure 19	Hand knitted mohair sock	73
Figure 20	Siirt blanket produced from mohair fibre	76
Figure 21	Cashmere goat	79
Figure 22	Prices of dehaired fine down (cashmere) fibres	86
Figure 23	The change in the price of Mongolian cashmere between 2006 and 2016	86
Figure 24	Cashmere fibre shedding in Raeini goat; a sequential and bilateral symmetrical fibre shedding initiates at the neck (on the left) and proceeds in a wave towards the rump (on the right)	88
Figure 25	Shearing (on the left) and dehairing by hand (on the right)	89
Figure 26	The cross-section, longitudinal view (fine down fibre and coarse guard hair) and the view of the scale layer (fine down fibre and coarse guard hair)	94
Figure 27	White and coloured cashmere fibres	102
Figure 28	"Mongolian Cashmere - made with" and "Mongolian Cashmere - pure" labels	109
Figure 29	Scarf and women's coat made of cashmere fibres	109

Figure 30	**Cashgora goat**	112
Figure 31	**Diameter ranges of mohair, cashmere and Cashgora fibres**	116
Figure 32	**Cashgora fibre, yarn and knitted fabric**	118

List of Tables

Table 1	Countries where luxury fibres of hair origin are produced, their properties, production quantities and prices	7
Table 2	Sources, places of production, properties, production quantities and raw yarn prices of luxury fibres of secretion origin	14
Table 3	Mohair production by country (million kg greasy)	28
Table 4	Classification of South African mohair	41
Table 5	Classification of American mohair	42
Table 6	Classification of standard and natural mohair	43
Table 7	A comparison of the single fibre tensile properties of wool and mohair	55
Table 8	Tenacity and elongation at break versus test length for mohair fibres and kemp hairs	56
Table 9	Changes in mohair strength and tenacity with goat age	56
Table 10	Moisture, grease and water-soluble matter content of raw wool and mohair	58
Table 11	The amino acid content (% mol) of mohair fibres	66
Table 12	Rules determined by IMA for using the mohair trademark (label)	71
Table 13	Average scouring yield of cashmere fibres of different origins	99

List of Tables

Table 14	Moisture, oil and water-soluble matter content of Chinese and Australian cashmere	100
Table 15	Price ($/kg) difference per quality in Mongolia	103
Table 16	Prices paid for dehaired cashmere by major UK processing company 1992-2002	103
Table 17	Amino acid composition (µmol/g) of cashmere fibres	106
Table 18	The amino acid (mol%) content of Cashgora fibres	117

Preface

It is today almost impossible to compete by producing simple and ordinary textile products with low added-value, the only way to survive is to focus on the production of high added-value products that require specialised know-how. From this point of view, it can be said that there are three main areas that offer important opportunities in textiles. The first of them are knowledge-intensive (not labor-intensive) textiles which require know-how, including technical, smart and functional textiles. The second area is the production of fashionable products. At this point, it can be said that the denim and piece garment sectors give big opportunity to produce innovative textiles. The third and final area is the niche product area covering low volume big value-added products. Luxury animal fibres, the subject of this book, fall within this area, since it offers important opportunities in terms of products that are light in weight but heavy in value. On the other hand, information on luxury fibres, especially their processing, is kept "like a secret" as stated in Prof. Dr. Lawrance Hunter's books. Although the number of publications on luxury fibres have increased in recent years, there is still very limited information available in this field.

Before starting the chapters of the book, I think it will be useful to give the definition and terminology of luxury fibres in this preface. Fibres such as mohair, cashmere, alpaca, vicuna, angora, silk are obtained from animals living in hard-to-reach regions of the world and are produced in very low quantities and therefore they are called *"Luxury Fibres"*. Luxury fibres give people wearing clothes made of these fibres a different status in society. This is confirmed by Fabio d'Angelantonio (CEO of the Italian luxury fashion brand Loro Piana, whose main materials are vicuna and cashmere) who said in an interview "If people buy Loro Piana, they enjoy feeling part of a private club." As B. Franck says in the preface of the book entitled "Silk, Mohair, Cashmere and Other Luxury Fibres", "A cashmere garment is the textile equivalent of a Rolls Royce, a diamond necklace or a holiday home in the Caribbean!". Although the word "luxury" evokes connotations such as being accessible by the limited population, beyond need, unnecessary, squander; Coco Chanel defines luxury as "a necessity that starts where the necessity ends". In addition, while luxury was limited to a more specific people segment in the past, it is increasingly getting accessible by wider population.

Like moving to a higher-level need after a need at a lower-level is met in Maslow's hierarchy of needs; for those who fulfill any kind of a need at a basic-level, meeting this need at a higher-level becomes a necessity, even if it may seem like luxury for those who cannot afford this need even at a basic-level. Such are the needs in the field of textile clothing. Throughout history, people used to dress only to cover themselves and thus to protect themselves from natural atmospheric conditions, and it was satisfactory enough for the clothes they wear to provide this. Later on, this became unsatisfactory, and they got into the expectation of wearing stylish and beautiful clothes and to dress for adornment beyond covering. Nowadays, people increasingly want their clothes to provide additional functions (like protection from various impacts (such as harmful microorganisms), notifying against changes in the body (such as fever), treating (such as relieving impaired circulation disorders with thermoregulatory effect) and so on) and/ or comfort and special touch, as well as covering and adornment. Clothings made of luxury animal fibres offer the highest level of features that a person can expect from clothing thanks to their unique softness and thermophysiological comfort. Pierre Giuseppe Alvigini expresses the passion that occurs in people who use luxury fibre as "It is not easy for someone who is used to drive Rolls Royce to start driving simple passenger cars, and for a person who uses luxury fibres, this becomes not a habit or a caprice, but simply a necessity.". The founder of Silk & Cashmere, Ayşen Zamanpur, expresses this passion on the website of her company as "If you touch the cashmere, it will touch you...".

In general, animal fibres other than wool obtained from sheep can be described as *"Luxury Fibre"*. Luxury fibres are also called *"Speciality Fibres"* as they are extraordinary fibres with very special properties not found in other fibres, or *"Rare Fibres"* as they are fibres grown only in certain regions of the world and produced in limited quantities. Meanwhile, Pierre Giuseppe Alvigini describes these fibres in his book as "Fibres Nearest to the Sky", which should be due to the fact that a significant portion of the animals from which they are obtained live in high places such as the Andes mountains. In fact, as William Shakespeare said in his Romeo and Juliet: "What's in a name? That which we call a rose, by any other name would smell as sweet.".

As is known, the group of animal fibres consist of natural and regenerated protein fibres. Luxury fibres, which are in the class of natural protein fibres, have two main classes: hair and secretion fibres. Since the hair fibres are produced by the follicle cells in the skin and they consist of keratin macromolecules, they are also referred to as *"Keratin Fibres"*. Hair protein fibres are generally called wool fibres. Although it is sufficient to call the fibres obtained from sheep just as wool, the fibres obtained from other animals are known with the name of the animal or with its private name. For example, the fibres obtained from Angora rabbit are called Angora rabbit wool or angora. The fibres of secretion origin are also referred to as *"Fibroin Fibres"* since they consist of fibroin macromolecules as the main component. Secretion protein fibres are generally called silk fibres. Although it is sufficient to call the fibres obtained from domestic silkworm (*Bombyx mori*) just as silk, the fibres obtained from other animals are referred to with the name of the animal (such as Tussah silk, Spider silk, Sea silk) or with their private names (Pinna fibres).

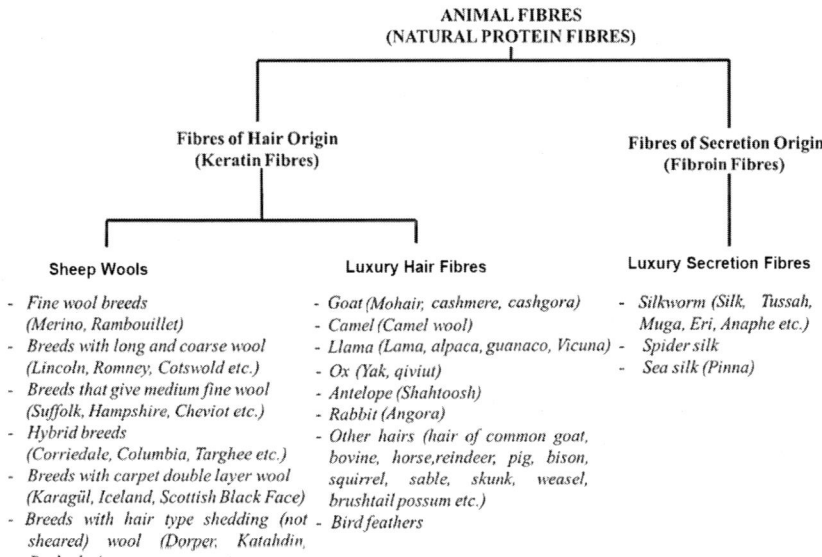

I started to get to know luxury fibres closely for the first time while preparing my undergraduate thesis entitled "Natural Fibres Other Than Cotton and Wool". Then during my PhD thesis on improving the dyeability of Angora goat (Mohair) and Angora rabbit (Angora) fibres, I had the opportunity to work experimentally. I later deepened my knowledge within the scope of various scientific research studies. Consequently, I came up with the idea of writing a book to convey my knowledge on luxury fibres and to contribute to the literature.

The subject of luxury fibres is a multidisciplinary issue including wide variety of expertise starting from the breeding of animals that these fibres are obtained from and then continuing with fibre harvesting and classification; dehairing process for fibres obtained from animals with bilayer fibres; and then transforming these fibres firstly into yarn, then surface (woven fabric, knitted fabric or nonwoven) and finally into clothing, and their finishing processes in various stages (fibre, yarn, fabric, garment, etc.). While the subjects such as the care and feeding of the animals that give luxury fibre, the cultivation of their characteristics through genetic breeding and the improvement of their various yield characteristics through crossbreeding are within the scope of veterinary and zootechnology disciplines; the production, classification, and examination of the various properties of luxury fibres are topics of interest to agriculture and textile engineerings. On the other hand, the processing of these fibres into yarn, fabric and textile materials, and their finishing processes is the main speciality of textile engineering. Considering the fact that the aforementioned issues are related and interacted to one and other, I can say that multidisciplinary studies in the field of luxury fibres may contribute even more to the development of this field.

Although I sometimes thought of collecting them in a single book, I believe that presenting luxury animal fibres, which are a bottomless pit in my opinion, in separate books will be a more "user-friendly" resource for you beneficiaries. In this book, after giving an overview of the luxury animal fibres, detailed information is given about the mohair, cashmere and cashgora fibres obtained from goats. Since it is out of my field of expertise, breeding of goats that give luxury fibre is excluded, and merely a brief presentation about each luxury fibre source goat is made at the introduction of each topic. Firstly, a general introduction is made about the animal that is the source of that fibre and the fibre it gives, and then information about historical background, world production, factors affecting production and yield, classification, microscopic properties, physical properties, chemical properties and end-uses of the fibre in question is given. It was mainly followed the way of explaining the general properties of luxury fibres, as much as possible rather than summarizing the literature even if in some cases it was done so.

I realised that I had a lot to say about luxury fibres, which are my passion, after I started writing the preface. Just like I said in one of my poems;

> like drops of pearl grains of a broken necklace
> my heart pours out the words inside of me one after the other

Although I have a lot more to tell, I think now it is time to end the preface without further ado.

I would like to end my words with my special thanks to Prof. Dr. Lawrance Hunter, co-author of this book and author of many other books and book chapters; and all other scientists who have contributed to the formation of my knowledge in this field with their research studies and books, and with the hopeful expectations of these books to be useful to anyone interested in luxury animal fibres, especially students.

Dr. Rıza ATAV, Prof.

Acknowledgements

We would like to acknowledge the publisher, editors, and support staff at Nova Science Publishers, they were very helpful during the various stages of developing and producing the volume.

Chapter 1

An Overview of Luxury Animal Fibres

Mohair, cashmere, camel, alpaca, llama, vicuna, guanaco, angora, yak, qiviut, silk, tussah, muga, eri etc. fibres are highly sought after and expensive due to their scarcity, and their certain unique and desirable properties which make these fibres very attractive. The origins of these fibres are as follows: mohair from the Angora goat; cashmere from the cashmere goat and camel, alpaca, llama, vicuna, guanaco from the respective animals belonging to the camelid family; angora from the Angora rabbit; silk, tussah, muga and eri from various silkworms. In summary, animal fibres, other than wool from a sheep, are described as "*Luxury Fibres*", also sometimes as "rare" or "speciality" fibres. The reasons fibres are described as "luxury" are their scarcity, generally being obtained from animals living in hard-to-reach areas of the world, and due to the fabrics obtained from them displaying distinctive features, such as softness and shine. They are quite expensive but have aesthetic appeal and give a different and important status to the people who wear them (Atav et al., 2003).

The use of luxury fibres in textiles began before recorded history in many civilizations (McGregor, 2018). Luxury is generally regarded as a sign of prosperity, power and social status (Gardetti and Muthu, 2015). Klaus Heine defines luxury as "a relative term that changes depending on who you ask, and indicates almost nothing" (Heine, 2012). Luxury varies according to cultural, economic and regional differences and preferences (Gardetti and Muthu, 2015). Robert H. Frank stated that in order to bring back the true values of life, the culture of "extremism" should be minimised (Frank, 1999). Environmental awareness has led many industries, especially in developed countries, to adopt more sustainable working methods, including the greater use of natural inputs and products. Natural products are not only technically valid components but can contribute to the value and pricing of the final product due to their superior environmental properties and compliance with socially responsible production and disposal requirements. The luxury, especially fashion, sector has a high environmental footprint and is responsible for a significant amount of waste. Therefore, there is an increasing demand for "sustainable luxury" (Karthik, Rathinamoorthy and Ganesan, 2015).

Globally, the livelihoods of many rural communities are based on fibre bearing animals, such as sheep, cashmere goats, alpaca and vicuna, and the importance of sustainable and humane husbandry techniques has increased. In fact, animal fibres are sustainable in that they are natural and biodegradable. Nevertheless, various associated issues, such as overgrazing and animal welfare, have become a matter of concern in terms of sustainability ("Sustainable Yarn", 2016). An example of a luxury fibre animal overgrazing problem is the cashmere goat. In the past century, the production of cashmere fibre has increased to such an extent that it has become unsustainable and posing a threat to the environment, with once beautiful pastures turning into deserts due to the increasing number of goats being bred for cashmere. This is having a devastating effect on the ecological balance of the planet (Karthik, Rathinamoorthy, and Ganesan, 2015). With respect to animal welfare, it is appropriate to mention the Angora rabbit. In 2013, animal rights organization PETA (People for the Ethical Treatment of Animals) released shocking classified footage involving certain angora farms, where frightened rabbits screech in pain and fear while their wool is being plucked, until only their bare, bleeding skin remains. Another fibre harvesting method involves shearing, using metal scissors (shears), with the front and hind legs of the rabbits being tied and/or stretched during shearing. Since the shearers work very quickly and the rabbits struggle, they frequently suffer deep and bleeding cuts on their skin ("Appalling Abuse of Rabbits", 2013). In the modern world, as in all production areas, the sustainability of animal fibre production has become an important requirement. Sustainable animal fibre production entails:

- Obtaining fibre from healthy animals that have been ethically raised and treated "humanely" throughout their lives,
- Raising animals in accordance with their natural instincts, social structures and needs,
- Growing the animals who have never been tied or tethered, never been treated with chemicals or hormones.
- Raising animals only on overgrown, uncultivated, and useless farmland grazing and processing their fibres without the use of harmful or undesirable chemicals or dyes and
- Harvesting fibre by manual combing (Karthik, Rathinamoorthy and Ganesan, 2015) or by careful shearing, which is painless and harmless for the animal.

Although wool, obtained from various sheep species, is by far the most widely used animal fibre, luxury animal fibres are also used in the production of clothing and other textile products. Such fibres are mostly used with sheep's wool (generically simply referred to as "wool") to facilitate processing and provide special effects, such as beauty, texture, colour, softness, flexibility, durability or lustre (McGregor, 2012). Although insignificant in terms of production volumes, such luxury fibres play an important role and are in great demand in the luxury and high value-added product market, especially the clothing and fashion market (Hunter, 2020). Together with wool, these rare animal fibres expand the range and aesthetics of textile products offered to consumers and the fashion industry (McGregor, 2018).

According to 2021 data, the world's total annual fibre production is approximately 113,000 million kg, of which approximately 29.5% (around 33,100 million kg) is made up of natural fibres, with sheep wool's share of total fibre production being around 1% at a production of approximately 1,000 million kg. The total production of protein fibres excluding wool, (i.e., luxury animal fibres), is 790 million kg, their share representing around 0.71% of total fibre production, approximately 173 million kg being silk fibres, which represents approximately 0.2% of the total fibre production. The total production of mohair, cashmere, camel, llama, alpaca, guanaco, vicuna, angora and Yak fibres is 570 million kg. representing around 0.24% of total fibre production ("Fibre Report", 2022). It is therefore clear that none of the luxury fibres is produced in large quantities relative to other fibres. This also applies to silk, with an annual production of approximately 173 million kg. Furthermore, the total annual production (approximately 790 million kg) of all luxury fibres is virtually negligible compared to the global production of textile fibres of approximately 113,000 million kg. It must be emphasised that the fibre production figures quoted for two-coated animals represent only the fine undercoat (down) and does not include the coarse outercoat or hair.

The production and harvesting of luxury fibres are both difficult and labour intensive. These fibres are usually grown in remote areas where access and transportation are difficult, and the prices of the fibres are high or very high. Since they are expensive, these fibres have a limited market, mostly involving wealthy consumers who purchase luxury goods, not only for their aesthetics (appearance, softness, warmth, handle, and comfort), but also because they are rare and expensive and associated with a certain status. Since the quantities produced are very limited, the price of all these luxury fibres is subject to large fluctuations. A sudden drop in demand from the market or a decrease in production can cause prices to decrease or increase by, for example, 50% within a few weeks. One of the most important issues to be met in the luxury fibre sector is to balance supply and demand. With respect to demand, there are basic statistical techniques currently used in the man-made fibre industry which can be used to provide a highly accurate forecast of demand for luxury animal fibres over a two-to-three-year period. With respect to supply, however, the problem is greater, since the supply of these luxury fibres is dependent on the number of animals available and environmental factors and cannot be increased or decreased very quickly. If accurate information about future demands could be transferred to producers by the relevant trade associations, they would have the opportunity to adjust their activities accordingly (Franck, 2001).

1.1. Hair Origin

The largest groups of hair-based luxury fibres are from goats and camelids (McGregor, 2012), namely Angora goat (*Capra hircus aegagrus*) and cashmere goat (*Capra hircus laniger*), two-humped Bactrian camel (*Camelus bactrianus*), as opposed to the Dromedary camel (*Camelus dromedarius*) which has one hump, and the South American camelid branch, llama, alpaca, guanaco and vicuña (Dalton and Franck, 2001). While it is generally agreed that the llama is a domesticated species of the guanaco genus, this is not the case regarding the genus to which alpaca and vicuna belong. Certain sources state that the genus Lama comprises four species: *Lama glama*, the *Llama* and *Lama pacos*, the *Alpaca* which are domesticated; and *Lama guanacoe*, the *guanaco* and *Lama vicugna,* the *vicuna* which are wild (Wardeh and Dawa, 2004). However, according to the more common literature the South American Camelids consist of four species, three, namely llama (*Lama glama*), alpaca (*Lama pacos*) and guanaco (*Lama hunchus or Lama guanicoe*), representing the genus Lama and the fourth, vicuna (*Vicugna vicugna*), being a separate genus Vicugna (Hunter, 2020). On the other hand, some studies showed that the alpaca was a domesticated vicuna and as such belonged to a different genus than the Lama supporting a reclassification as *Vicugna pacos* (Kadwell et al., 2001; Marin et al., 2007).

Like the Angora goat (*Capra hircus aegagrus*) and cashmere goat (*Capra hircus laniger*) representing the genus Capra, other hair-based luxury fibres from the Caprinae sub-family are the Tibetan antelope (*Pantholops hodgsoni*) representing the genus Pantholops, and Musk ox (*Ovibos moschatus*) representing the genus Ovibos. Although the Musk ox visually resemble some kind of ox or bison, they belong to the Caprinae sub-family, making them more closely related to sheep and goats than cows or bison (Muskox, 2023). Nevertheless, since it is commonly known as Musk ox, it will be included here under the heading of ox.

Beyond the goat (Caprinae) and camelid (Camelidae) families, there are animals belonging to two other families from which hair-based luxury fibres are obtained. They are the Tibetan ox (Yak) (*Bos grunniens*), belonging to the bovine family (Bovinae); and Angora rabbit (*Oryctolagus cuniculus*), belonging to the rabbit family (Leporidae).

There is also a certain amount of scientific interest in the possibility and advantage of interbreeding between the different breeds within both the goat and camelid groups. Examples include the huarizo and misti crosses between the alpaca and llama species and more recently Cashgora (Dalton and Franck, 2001). In Figure 1, main sources of luxury animal fibres are given.

Apart from the luxury fibres given in Figure 1, there are also some special fibres of hair origin from various animals (goat hair, cattle hair, bison hair, horsehair, reindeer hair, beaver hair, opossum hair, etc.) and bird feathers (duck, goose, etc.). The relatively coarse hairs obtained from the above-mentioned animals are generally used in saddlebags, tents, sacks, etc., while bird feathers are used as a filler in pillows, quilts, etc. Although coarse hairs are used in the production of various textile products, since they do not produce high value-added clothing items and at the same time there is no significant trade in them in the world markets, they are not considered as luxury fibres, but as special (or speciality) fibres. However, very small amounts of fine fibres obtained from some of these animals (e.g., opossums) are blended with wool and used in limited quantities to produce very special textile garments.

Table 1 lists the countries where luxury fibres of hair origin are produced, their properties, production quantities and prices.

Table 1 excludes Shahtoosh fibres obtained from the Tibetan antelope found in the Tibet region of China. These animals are usually killed to obtain their fibres, representing one of the highest quality fibres and which is used in the production of "ring scarves (so named because such scarves can pass through a wedding ring)" (Hunter, 2020). It is estimated that around 20,000 Tibetan antelope are annually killed for their fibre which varies between 9-12 µm and with 120-150 g fibre being obtained from an animal per year.

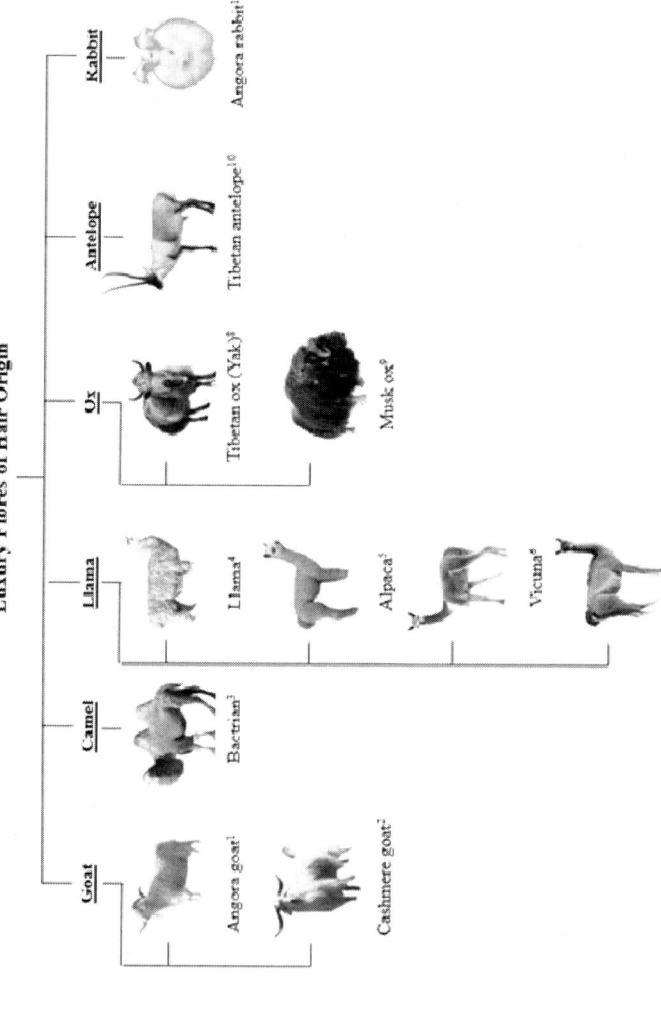

Figure 1. Main sources of luxury animal fibres (adapted from 1: "Angora Goat", 2008; 2: Wallace, 2008; 3: "Pascuali Filati", 2019; 4: "Llama", 2020; 5: "Alpaca", 2010; 6: Gallo, 2016; 7: "Guanaco", 2020; 8: "Domestic Yak", 2020; 9: Nosowitz, 2016; 10: Naresh, 2018; 11: "Angora Rabbit", 2020).

Table 1. Countries where luxury fibres of hair origin are produced, their properties, production quantities and prices

Fibre Name	Country of Manufacture[a]	Fibre diameter (μm)[b]	Fibre length (mm)[b]	Fibre production (kg/year)[b]	Production (ton/year)[b]	Price ($/kg)[a]
Mohair	South Africa, USA, Turkey, Argentina, Lesotho, Australia, Russia, Central Asian Countries	22-40	50-110	2-6	5,000	4-30
Cashmere	China, Mongolia, Iran, Afghanistan and Other Asian Countries, Russia and Australia	13-19	20-50	0.1-0.4	10,000	35-70
Cashgora	Central Asian Countries and Russia	18-23	30-60	2-4	50	8-20
Camel wool	China, Mongolia	15-25	30-60	2-5	1,000	10-12
Alpaca	Peru, Chile, Argentina, Australia, USA, EU Countries	18-35	60-80	2-4	7,000	5-20
Llama	Bolivia, Peru, Argentina	20-32	50-80	1.1-3.5	1,000	5-20
Vicuna	Peru, Chile	11-15	20-40	0.2	5	200
Guanaco	Argentina	14-20	20-50	0.3-0.9	10	50
Yak	Tibet, China, Mongolia	15-25	25-50	0.2-1.3	7,000	15
Qiviut	Greenland, Canada, Norway, Alaska	13-20	40-80	2-3	3	-
Angora	China, Chile, South Africa, France	10-18	30-60	0.5-1	4,000	20-30

(a: McGregor, 2012; b: Hunter, 2020)

Raw Shahtoosh prices range from $1,500 to $2,000/kg. Shahtoosh shawls are usually priced between $2,000-8,000 sometimes even up to $15,000 depending on size and quality, ("Chiru Facts", 2011). Until a ban was introduced in June 2000, the states of Jammu and Kashmir were the only places in the world where Shahtoosh fibres could be produced legally. In 2002, a global ban on Shahtoosh sales was introduced ("Shahtoosh and Chirus", 2012).

The production and price of luxury fibres are constantly changing, depending upon demand. While the demand for these fibres and their prices increased in the 1980s, the impact of the economic crises experienced in the 1990s decreased the demand for these fibres. Furthermore, China became the leading country among the luxury fibre producing countries, as a result of the great social and economic regime change in China in the 1980s (Atav et al., 2003). In terms of production, cashmere is the largest (Table 1), followed by alpaca, yak and mohair. Excluding the qiviut fibre, which is the down fine fibre obtained from musk ox, the most expensive fibre is vicuna, followed by guanaco and cashmere. Fabrics made from luxury hair fibres are generally much more expensive than those from wool. For example, a cashmere sweater is sold for around 4 times (£100) more than a wool sweater equivalent. A fabric containing guanaco fibre costs around £500 per metre (Atav et al., 2003). On the other hand, a scarf made from vicuna fibres, the world's most expensive fibre used in the manufacture of apparel, can cost around $1,500, while a vicuna sports jacket can cost $21,000 ("Vicuña", 2020).

All luxury fibre animals, except the Angora rabbit, live in regions where climatic conditions are harsh, typically ranging in temperature from well below freezing at night to warm or in some cases tropical during the day (Dalton and Franck, 2001). The nature of the animal's environment also largely determines the natural colour of the hide and hair for camouflage and other purposes. Due to the extreme temperatures they encounter, most of these animals have developed two different layers of fibre (i.e., are double coated), an outer layer (outer coat) of coarse, medullary protective hair produced by primary follicles that provide protection from sun, rain and dust, and a down or inner layer (inner coat) of finer, shorter fibres produced by secondary follicles that provide outstanding insulation against extreme temperatures. Exceptions to this are the Angora goat and alpaca, which, like sheep, have essentially a single layer of fibre, but which still contains a combination of primary and secondary follicle fibres. In some cases (e.g., llama), a group of intermediate fibres may be present in addition to the fine down fibres and coarse outer hairs. Most luxury fibres tend to be finer, less curly, and smoother than wool. For a given species, the younger the animal and the higher the altitude, the finer the fibre in general (Hunter, 2020). In Figure 2, the fineness ranges of various luxury animal fibres are compared.

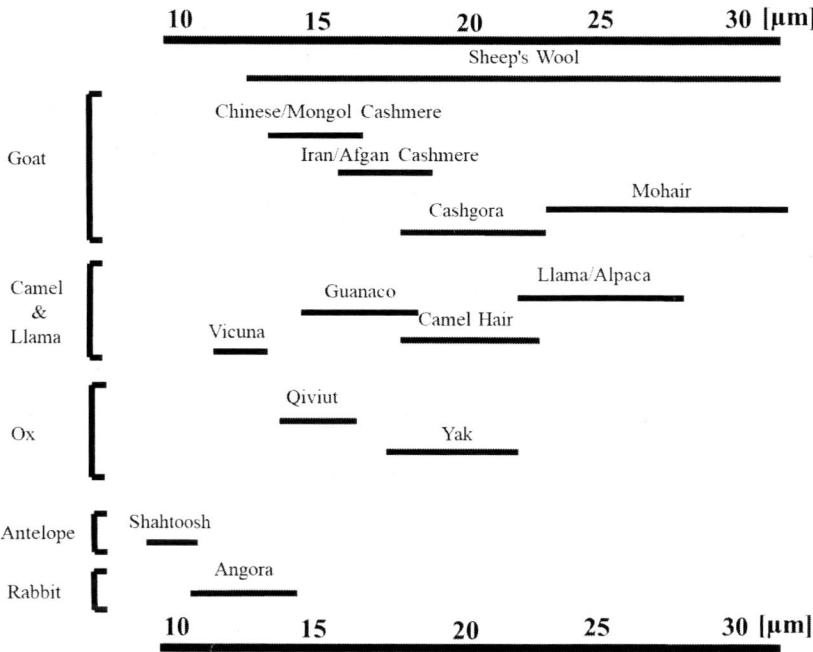

Figure 2. Fineness (diameter) ranges of various luxury animal fibres (modified from: Phan, 2007).

As can be seen from Figure 2, most luxury fibres are finer than wool. Among the luxury fibres, Shahtoosh, vicuna and angora fibres are generally the finest, with mohair, llama and alpaca fibres generally the coarsest. In terms of mean fibre length, most luxury fibres are generally 2 to 6 cm (20 to 60 mm) long, although mohair generally varies between about 5 and 11 cm (50 and 110 mm). Harvesting fibre from various animals is carried out by various methods, such as shearing, combing and plucking, as well as collecting the fibres that are shed during the moulting season.

Chemically, luxury fibres belong to the same protein (keratin) family as wool, although their morphological structure and surface structure differ, and they are often medullated. Due to their similar chemical compositions, they are not easy to distinguish from each other chemically (Hunter, 2020).

Although substantial advances have been made using modern techniques, luxury fibres are often difficult to distinguish reliably from one another. For example, the quantification of the components of a blend fabric containing wool and cashmere or mohair depends on the skill and experience of the operators and remains expensive, time consuming and somewhat even a little subjective (Dalton and Franck, 2001).

In most luxury fibres, the scale layer is less pronounced (smoother or finer) than in wool. Compared to wool, which usually has scale heights of 0.6 μm and higher (typically about 0.8 μm), the scale thickness (height) of luxury fibres is around 0.4 μm or even finer (usually 0.3 to 0.4 μm), even if the scale height for mohair can be as high as 0.5 μm or even sometimes slightly higher. In addition, the scales in luxury fibres are generally more widely spaced than in wool. The average scale frequency per 100 μm ranges from about 4 for mohair to about 12 for vicuna. The ellipticity of the luxury fibres ranges from about 1 (i.e., fully circular) to about 2, with an average of about 1.2 to 1.3 (Hunter, 2020). A key difference from wool is that many of these fibres tend to be medullated and have microscopic air spaces within the fibre structure. This makes the fibre lighter and contributes to its insulating properties (Dalton and Franck, 2001).

The fine down fibres (undercoat) of double coated animals, which are usually shed in spring, are the most valuable for textiles, particularly for apparel, due to their fineness, softness, lightness and good thermal insulation properties, and these need to be manually or mechanically separated from the unwanted coarse protective (outer coat or guard) hairs (Hunter, 2020). In many cases, medulla-containing fibres are relatively coarse and they show up lighter, even white, in dyed textile materials as they reflect and refract light differently compared to the other solid fibres. However, in some fibres such as angora, medullation is a desirable fibre characteristic and an important feature of the finished product, since it provides a soft and fluffy appearance (McGregor, 2018), with excellent heat (thermal) insulation.

Separation of the fine down fibres from the coarse guard hairs (i.e., dehairing) is performed before combing or carding. Exactly how each top or yarn manufacturer does this is still relatively secret and propriety knowledge to each company (Dalton and Franck, 2001). The finer the fibre and the lower the percentage of coarse hair (protective hair) after dehairing (i.e., after the separation process), the better the textile quality and value of the fibre. In this regard, successful separation largely depends upon the differences in fibre diameter, stiffness, linear density, friction, length, and inter-fibre cohesion between the two components (Hunter, 2020).

Apart from the separation process, the spinning of these luxury animal fibres is carried out according to the woollen process for shorter fibres and the worsted process for longer fibres. However, all these fibres have smoother surfaces than wool because the scales of the fibres are less pronounced and more widely spaced (Dalton and Franck, 2001). For those fibres with crimp, the crimp is usually not as pronounced or of such high frequency as for fine wool, and in some cases (e.g., mohair), words like ringlet or waviness better describe the type of crimp (Hunter, 2020). Since the fibre cohesion and friction of most luxury fibres are lower than that of wool, special precautions and conditions and/or blending with other fibres, such as wool, are required to obtain acceptable processing performance and yarn quality. Although the processing is mostly done on machines similar to that used for wool, the machines and processing conditions, settings etc. must be adapted and optimised according to the specific requirements of each of these fibres. Such information is usually kept secret by each of the manufacturing companies in this field (Dalton and Franck, 2001).

All luxury animal fibres are keratin fibres, their chemical and physical structures being similar to those of wool. Since their chemical structure is very similar to that of wool, the dyeing techniques, machinery and dyes used in dyeing these fibres are similar to those used for wool. Nevertheless, special, and mostly milder, conditions are often required during the dyeing and finishing process of these fibres, e.g., processing time, temperature and pH, so as to preserve their specific and desirable properties, such as lustre, and to minimise damage (Hunter, 2020). Furthermore, since the smoother surfaces of these fibres reflect light differently to wool, it may be necessary to modify the dye recipes to obtain the same specific hue (Dalton and Franck, 2001). Scouring, processing, dyeing, finishing, knitting and weaving of luxury animal fibres globally are done by specialised companies within the wool textile industry (McGregor, 2012).

World textile production and consumption continue to increase due to increasing global populations and wealth and new and diverse applications of textiles. Many new synthetic fibres have been developed to overcome certain weaknesses or deficiencies in natural fibres and increasing technical and functional requirements. Therefore, animal fibres are becoming more appealing to smaller niche markets where they are able to compete with other fibres on the basis of their specific features and strengths. Hence, it is essential to have the latest knowledge and technology available to produce the premium quality associated with such rare animal fibres (McGregor, 2012). It is also important to note the recent development of new textile products, which capitalise on both the technical advantages (especially low flammability) and superior image, aesthetics and comfort of these protein fibres (Dalton and Franck, 2001).

1.2. Secretion Origin

Although fibre (filament) is secreted by many insects in the world, the number exploited economically is quite low (Yazıcıoğlu and Gülümser, 1993), the silkworm family being the exception and the source of a luxury fibre. Silk fibres, referred to as the "Queen" of Fibres, secreted by silkworms, have been used in textiles for some 5000 years, their main desirable features include lustre, handle, strength, extension, durability, comfort and dyeability.

In order to pass one of the life stages of the silkworm, the secretion, coming from 2 silk glands which combines in a hole at the tip of the lower lip, comes out as a single (composite) silk thread (filament), which is then formed into a cocoon surrounding the silkworm (Tarakçıoğlu, 1983). Although there are silk fibres, such as tussah (*Antheraea mylitta*), muga (*Antheraea asamsa*), eri (*Phylosomia ricini*), anaphe, which are obtained from various wild silkworms, today silk, which has gained importance in terms of production-consumption amounts around the world, is obtained from the domesticated silkworm of *Bombyx mori*. Although the mulberry fed *Bombyx mori* species can be domesticated, others are generally difficult to domesticate, living in colonies, feeding mostly on the leaves of some forest trees. For this reason, these silkworms are called "wild silkworms" and their silk "wild silk". Silkworms, which do not feed on mulberry leaves but feed on the leaves of other trees, are generally included in the *Antheraea* and *Phylosomia* genus of the Saturniidae family. In addition, fibre is obtained from some Anaphe species belonging to the *Thaumetopolidae* family (Yazıcıoğlu and Gülümser, 1993). Figure 3 shows main sources of luxury secreted fibres.

Apart from the "silk type" fibres given in Figure 3, there are also some other special such fibres, including spider silk obtained from various spiders, especially *Nephilla clavipes* species, and sea silk (pinna) fibres obtained from sea mollusks *Pinna nobiles* or similar marine animals (Yazıcıoğlu and Gülümser, 1993). The most productive silk producers are insects in the orders Araneae (spiders) and Lepidoptera (lepidoptera). Spiders produce multiple types of silk in very small quantities every day throughout their lives.

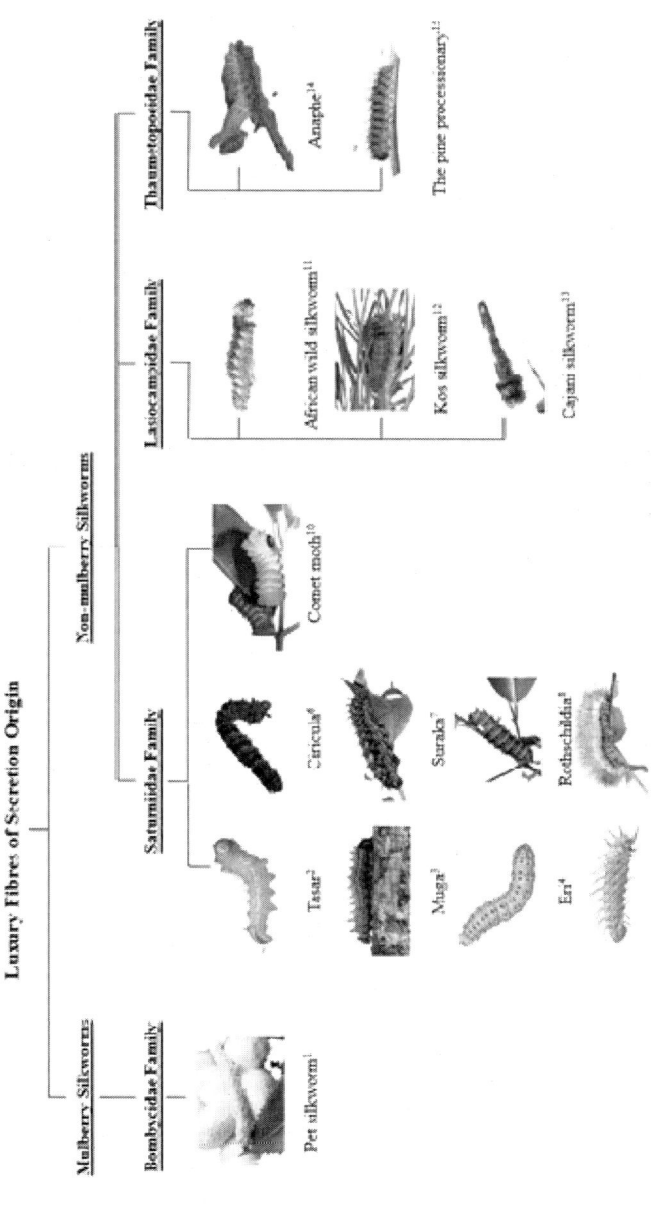

Figure 3. Main sources of secreted luxury fibres (adapted from 1: Bohanec, 2017; 2: "Chinese Oak Silkmoth", 2020; 3: "Muga Silkworm", 2020, 4: "Samia Cynthia Drury", 2020; 5: "Giant Silkworm", 2011; 6: "Cricula Trifenestrata", 2020; 7: Hertog, 2020; 8: Huggins, Sookdeo and Cock, 2018; 9: Krejčík, 2011; 10: Albaugh, 2020; 11: "Wild Silk", 2020; 12: Ziegler, 2007; 13: Cox, 2009; 14: Mbahin et al., 2010; 15: Manzano, 2020).

Some Lepidoptera can also produce small amounts of silk daily to mark their foraging paths (Craig, Weber and Akai, 2012). Spider silk, which is stronger and more flexible than steel wire (on a linear density basis), could not be produced in large quantities for commercial use until today (İlleez, Öktem and Seventekin, 2003).

Sea silk is an extremely fine, rare and valuable fibre being produced from long silky filaments secreted by a gland on the bottom of several bivalve mollusks (especially *Pinna nobilis L.*), with which they attach themselves to the seabed ("Sea Silk", 2020). Weaving and embroidering with sea silk is a long and laborious process, and only an Italian woman, named Chiara Vigo, can apparently do this today. Her products are exhibited in various museums ("Sea Silk Weaver", 2017). Although both spider silk and sea silk fibres are very valuable, they are rarely traded on the world markets, and therefore not considered as luxury fibres, but rather as special fibres.

Table 2. Sources, places of production, properties, production quantities and raw yarn prices of luxury fibres of secretion origin (Tarakçıoğlu, 1983; Yazıcıoğlu and Gülümser, 1993; Currie, 2001; Atav and Demir, 2009; "Fibre Report", 2022)

Fibre Name	Producing Countries	Fibre fineness (denier)	Filament Length (m)	Production (ton/year)	Price ($/kg)
Silk	China, India, Japan, Brazil, Thailand, South Korea, Vietnam, Uzbekistan, Iran	1-1.5	800-1,200	173,000	20-22
Tussah	China, Japan, India	5-10	800-1,500	3,000	250-300
Muga	India	4.5	350-400		500-700
Eri	India and other Far East countries	5-10	Staple fibre		50-60

Information on the general properties, production quantities and prices of various luxury secreted fibres are given in Table 2.

In Table 2, anaphe silk, obtained from some anaphe species belonging to the Thaumetopolidae family in Uganda and surrounding African countries, is not included (Yazıcıoğlu and Gülümser, 1993). In fact, African wild silk has not been able to enter the textile industry so far, although it has been proven by many tests by theorists and practitioners that African wild silk has many important advantages over Asian wild silk ("Silk", 2008). When the production of the fibres in Table 2 are compared, it is clear that by far the biggest share belongs to the silk obtained from the domestic silkworm. In terms of price, muga fibres, which have a natural golden yellow colour and are called "golden silk", are the most expensive.

Wild silkworms, producing tussah, muga and eri fibres, are important for basic entomological and biotechnological research in various countries where science and technology have advanced. When it comes to exploiting wild silk, technological research and development are limited to China, India and Japan in Asia. India is the world's largest tussah silk producer after China, as well as the world's largest producer of muga silk, with also a great production potential for eri silk. Since the fibres obtained from these silkworms have high added value, they generate significant income in the countries where they are produced. For this reason, attempts continue to be made to increase production in various countries, such as India and China, where these silkworms occur. Although the usage areas are limited to traditional clothes and special designs, it is thought that, if production can be increased, the market will expand as the prices will decrease to more affordable levels (Atav and Demir, 2009).

A *B. mori* eggs hatch in 10 days and then develop into larvae, (silkworm caterpillars). Silkworm larvae eat mulberry leaves almost non-stop for 35 days, increasing their weight 10,000 times from a tiny speck to a fat worm (Karthik, Rathinamoorthy, and Ganesan, 2015). The silkworm starts to make cocoons 35-45 days after it emerges from the egg and finishes it within 4-5 days. If the wet cocoon is left in this way, the caterpillar (chrysalis) turns into a moth (butterfly) after 14-16 days and pierces the cocoon and emerges. Therefore, it is necessary to kill the chrysalis in the cocoon, except for those reserved for egg production. On the other hand, it is necessary to dry the cocoon to reduce the risk of mould and damage by fungi (Tarakçıoğlu, 1983).

A silk fibre consists of two fibroin filaments surrounded by an egg white substance called sericin (Tarakçıoğlu, 1983), both fibroin and sericin being protein. Silk fibres, which have a different structure from other keratin-based protein fibres, consist of approximately 75% fibroin and 25% sericin (Yurdakul and Atav, 2006). Most of the amino acids in wool are also found in fibroin. However, the ratios of the various amino acids in fibroin vary according to the type of silkworm, nutrition and environmental conditions (Yazıcıoğlu and Gülümser, 1993). Unlike keratin, which has a α-helix structure, fibroin has a β-structure. As a result, there are more H-bridges between the macromolecules. This causes silk to have a higher tensile strength than keratin-based fibres, but a lower extension at break (Tarakçıoğlu, 1983). There are also differences in the tensile strength and elasticity of the different types of silk *B. mori* silk is more elastic than the silk produced by the Anaphe species but has a similar tensile strength. On the other hand, compared to the silk produced by the Saturniidae family (Tussah, muga and eri), it has a lower elasticity but higher breaking strength (Craig, Weber, and Akai, 2012).

Cocoons need to be treated with hot water and steam in order to find and pull (reel) the silk fibres from the cocoons, and this process is called "cooking" or "boiling the cocoon". After that, the reeling (filature) process takes place in which the silk filaments are unwound from the cocoon. Reeling is the winding of the silk threads coming from a certain number of cocoons to obtain the desired number and linear density of the threads. However, not all the silk can be reeled as filaments due to the cocoon being damaged for example. The silk part that cannot be reeled is called "noil silk" or "waste silk". After a sericin removal process, they are separated into staple (shorter) fibres and spun into yarn using a staple spinning process like worsted spinning, the yarn being called "schappe silk". While producing schappe yarn, the noils are not discarded but are spun into lowest-quality silk yarns, called "bourette yarn" (Tarakçıoğlu, 1983).

Eri silkworm cocoons consist of short protein fibre (filament) segments. Therefore, it must be spun into staple yarn, like cotton or wool. It is also the only silk fibre that can be reeled without killing the chrysalis. For this reason, Eri silk is also referred to as "peace silk" and "strict vegetarian (vegan) silk" (Atav and Demir, 2009). Anaphe fibres are too weak to be reeled from the cocoon as a single filament without breaking (Agbadudu and Ogunrin, 2006). For this reason, Anaphe fibres are also considered as "schappe silk", like noil silk.

The finishing process of silk fibres begins with the partial or complete removal of the sericin. Silk fibres that have almost all of the sericin removed, that is, consisting only of fibroin, are called "cuite silk". Silk, from which approximately half of the sericin has been removed, is called "souple silk", and the silk from which only a very small amount (3-5%) of sericin has been removed is called "ecru silk". Apart from sericin removal, bleaching as a pre-treatment can be done with reducing or oxidizing agents, as in the case of wool (Tarakçıoğlu, 1983). Although the dyeing properties of all protein fibres are basically the same, the absence of surface scales in silk fibres and the fact that they are more resistant to alkali than wool, form the basis of the difference in the dyeing properties of these two fibres. Silk fibres can be dyed with some direct dyes used in dyeing cellulose fibres as well as the dyes used in dyeing wool, excluding 1:1 metal complex dyes (Yurdakul and Atav, 2006). After dyeing, a finishing process is applied to give silk products the desired handle.

Another silk-specific finishing process is the charging (weighting) process with tin, phosphate and silicate salts and/or vegetable substances, such as tannin and gall oak extract, in order to compensate for the weight loss resulting from the sericin removal and sometimes even to increase its weight to more than in its raw state (Tarakçıoğlu, 1983).

Silk, often referred to as the "Queen of Fibres", is gaining increasing popularity worldwide (Atav and Namırt, 2011). The word silk has several connotations for people. For example, when a lady is asked what the word "silk" means to her, she will probably reply with words like "sensitivity", "luxury", "charm". Silk has been a fibre that has inspired writers and poets throughout the ages. In addition, silk has entered the daily language with idioms such as "smooth like silk" and "silky hair". Silk even features in fields other than textiles, for example in promoting various products and services, such as cigarettes, shampoos and airlines (Currie, 2001).

Fabrics woven from pure silk are used in the making of super quality clothing all over the world. Silk fabrics are the first fabrics that come to mind, especially in women's outerwear and underwear, scarves and shawls, men's shirts, ties and similar luxury clothing. Clothing made of silk fabrics are widely used in Japan, including in traditional "Kimono" clothing (Atav, 2002). Tasar and muga fibres are also used to make various traditional garments in India. Stylish shawl and scarf making is popular because of the thermal properties of these fibres. Eri fibres can be blended with cotton, wool, jute or mulberry silk to create exotic fabrics that can be used in jackets or sweaters or in furniture and interior decoration (Atav and Demir, 2009). Anaphe silk, on the other hand, is used together, with cotton, in the production of a fabric called "Soyan" by the locals in Southern Nigeria (Tortora and Johnson, 2013).

Chapter 2

Luxury Fibres Obtained from the Goats

The goat was domesticated between 6,000 and 7,000 BC, the origin of domestic goat breeds being the *Capra prisca* goat in the eastern region of Central Europe, the *Capra falconeri* in Afghanistan, and the *Capra aegagrus* wild goat breeds living in the mountainous regions of Anatolia and Iran. Goat breeds are classified as follows (Günlü ve Alaşahan, 2010):

a) Dairy breeds (Saanen, Toggenburg, Malta, Aleppo, White German, Nubian and Kilis goats),
b) Meat breeds (Boer, Jamnapari and Black Bengal goats),
c) Fibre/hair breeds (Mohair (Ankara) and Cashmere goats)
d) Dual breeds (hair goat and Sudanese goat)
e) Fur and leather (skin) breeds (Marradi and Nubende goats)

Another example, which is included in the group that produces fibre, other than mohair and cashmere, is the Cashgora. Cashgora is defined as a hybrid fibre (Bolat, 2006). The Cashgora goat is obtained by crossing an Angora goat with a Feral Cashmere type goat ("Goat Fibres", 2010).

There are essentially two main types of hair/fibre producing goats, namely those with two coats (e.g., Cashmere type), having an outercoat of coarse guard hair and an undercoat of fine fibre (down), and those with a single coat (e.g., the Angora and Cashgora goats). The coarse fibres or hair (i.e., outer coat or guard hair) are produced by the primary follicles and the fine down fibres (undercoat) by the secondary follicles. The most valuable of these are the fine down fibres, or more specifically cashmere type fibres, since even if it is used in a very low proportion (e.g., 2.5%) in a fabric, it can significantly change its handle and softness as well as its image. Although the term "cashmere" is defined as fibres obtained from the Cashmere goat, it is generally also used for down fibres finer than 19 μm obtained from goat species other than the Angora and Cashgora goats (Göktepe, Canipek and Soysal, 2018).

If we leave aside the coarser guard hairs that are still mainly used in tent weaving by nomadic tribes, commercially produced goat fibres are essentially limited to mohair from Angora (also sometimes called Ankara) goats and "cashmere" from Cashmere goats (van der Westhuysen, 2005).

2.1. Mohair Fibres

The hair, called Mohair, produced by the Angora goat (Figure 4), which has spread from Turkey to many other parts of the world, is long, white, fine and lustrous (silk-like) (Öktem and Atav, 2007). The word Mohair is derived from the Arabic word "Mukhayar" (also spelled as Makhayar and Mukhaya), which means "the best of elite fleece", "exclusive selection", "silky goatskin fabric", "bright goat hair fabric", "hair fabric" (Hunter, 1993). In addition to these, the term *"Diamond Fibre"* is also used for mohair, due to its silky lustrous (glossy) appearance (Öktem and Atav, 2007). Mohair is also called as "Filik" in Turkey colloquially. The international abbreviation for mohair fibre is WM (Wool Mohair) ("Fabric Abbreviations" 2020), which means "Mohair Wool". Mohair is considered one of the most important speciality (luxury or rare) animal fibres and has the highest production of all specialty animal fibres (Hunter, 1993). The Angora goat (*Capra hircus aegagrus*) belongs to the *"Capra hircus"* species, which includes European dairy goat breeds and all other domestic goat breeds (hair goats) and is closely related to Asian Cashmere goats and certain Himalayan goat breeds (Hunter and Hunter, 2001).

The live weight of the Angora goat (Figure 4), which has a small to medium body structure, is 30-40 kg for ewes and 45-55 kg for bucks. Their heights are around 50cm for does and 65cm for bucks. They have ears of medium length which are partially drooping. Does have short, backward curved and straight horns, while bucks have corkscrew-like horns extending backwards and laterally(sideways) (Şengonca and Koşum, 2005). The fecundity varies from 60 to 100% (89% on average) for Angora goats, in relation to the breeding and selection of the flock. Angora goats don't give much milk and the kids are generally weaned when they are two months old (Alvigini, 1979).

Angora goats can withstand extreme temperatures, except that after shearing they are very sensitive to cold, and especially to a combination of cold, wind and/or rain (Hunter, 2020). These goats normally live in areas with low rainfall and low humidity (Hunter and Hunter, 2001). The areas where Angora goats are successfully bred include steppe regions at an altitude of 450-1500 m above sea level, with low humidity and rainfall and small shrubs.

These goats can benefit from the vegetation on very inclined slopes while other farm animals cannot (Vatansever, 2004), and can be raised on many different types of pasture. Angora goats are much more efficient than sheep in converting feed into fibre. Sheep, on the other hand, are more effective at converting nutrients into body mass (Hunter, 2020).

Figure 4. Angora (Mohair) goat (courtesy of Gerda Hayward – The livestock photographer).

Although Angora goats are primarily bred for their mohair (ie, fibre or hair) yield, their meat, milk and leather are also used, their meat being considered delicious and healthy. The fact that the meat of hair goats does not have the "heavy goat odour" makes Angora goats preferable. Angora goats do not produce much milk, they only produce enough to feed their kid. The skin of Angora goats is thin and soft and is used for making hides, suede, shoes, bags and gloves ("Mohair and Sof", 2018).

The Angora goat is recognised as unique among goats, since its fibres produced by the primary and secondary follicles do not differ significantly, and it produces essentially a single coat. For centuries, mohair has been regarded as one of the most luxurious and best quality fibres available. It is usually a long, straight (not crimped, but usually wavy), smooth and lustrous fibre, white or off white in colour, and can be dyed to both dark and bright colours (Hunter, 2020). The characteristic features of mohair are excellent lustre, strength, elasticity, resilience, abrasion resistance, resistant to staining and soiling and good drape, moisture and sweat (perspiration) absorption and permeability, insulation, comfort, pleasant handle and a relatively low flammability, felting and pilling tendency (Hunter and Hunter, 2001; Hunter, 2020).

2.1.1. Historical Background of Mohair

Although the exact origin of the Angora goat is unknown, its origin is believed to be the Asian Himalayas or the Tibetan Plateau. Later, they came to Ankara (known as Ancyra in ancient times), which was the capital of Phrygia in Anatolia and from which the name Angora was derived. Accordingly, the Angora goat appeared in Turkey after the Middle Ages (at least in the thirteenth or fourteenth century). There are traces of the existence of Angora goats in old records of up to the 11th, 12th, and even 14th centuries B.C. In the book of Exodus, 1500 years before Christ, it is reported that the Israelites, who fled from Egypt, took the goats with them, from which they wove cloth from fibre (mohair) to be used as coverings on altars and curtains in temples (Hunter, 1993; Hunter and Hunter, 2001).

Raising Angora goats in Ankara started after a long and difficult journey of thousands of kilometres from Turkistan to Anatolia. The journey started in the 13th century when Genghis Khan drove Suleiman Shah and his people out of the Turkmen lands. Suleiman Shah drove his herd of goats short distances every day, crossing the Caspian Sea and finally reaching the Euphrates River. However, he drowned while trying to cross the river and his son Ertuğrul then took over. Ertuğrul arrived in Konya and became a subject of Sultan Alaaddin, for his commendable services he was rewarded with a land (dormitory) stretching from Kayseri to Ankara. After the Turks settled in Anatolia, the goats adapted to the climatic conditions of Central Anatolia, their breed characteristics becoming established and clear, and their reputation as a special breed specific to this region spreading to other parts of the world. The Angora goat got its name from Ankara ("Mohair Association", 2020).

For centuries, Angora goats were bred only in Anatolia, mainly in Ankara and surrounding regions, the first fibre (mohair) obtained from this goat remained under the monopoly of this region. Starting from the 15th century, a strong mohair-based weaving industry developed in Ankara, and the fabrics obtained from mohair became popular both in the country and abroad. It was the most important industry and source of income for Ankara and as a result the city experienced a very active commercial and social life ("Mohair and Sof", 2018).

The fact that the superior quality of mohair was noticed by the Europeans who bought mohair products, and that mohair was in great demand, lead to the idea of bringing Angora goats to Europe ("Mohair and Sof", 2018). The versatility and beauty of this spiral horned goat and its fibre (mohair) were discovered by a Dutchman in 1550, initiating a demand and development which formed the beginning of the mohair industry in Europe. Four years later, a pair of Angora goats was sent to the Holy Roman Empire as heraldic gifts ("Mohair Association", 2020).

The British came to Ankara for trade purposes after 1580, and in 1583, Queen Elizabeth I gave permission to 12 English merchants to trade in tablecloths. According to the records, these traders established a company called the Levant Company (in 1590). Actually, the Dutch first started trading with the English, then they competed with them, with the Polish merchants being the most active ("Mohair and Sof", 2018).

Evliya Çelebi, who had been in Ankara during the 1640s, spoke of the city of Ankara as follows: "This is the place of soft mohair (fabric made of mohair fibre). It is unlikely to exist anywhere else on earth.". Then he described the Angora goat as follows: "The Angora goat is like white milk, and maybe there is no white creature like it, soft yarn being made from its wool. If they cut the fleece of these goats with scissors, the yarn will be tough. But if they pluck it, it will be soft like the silk of Prophet Job. The wool (hair) from both male and female goats are soft." ("Mohair Association", 2020). Although it is not known exactly when manufacturing of mohair products in Ankara and its surroundings started, it appears to have commenced with the arrival of the Angora goat in Anatolia. It is said that in 1655, there were 13,555 looms around Ankara and that 20 thousand rolls of fabric were exported to Europe ("Mohair and Sof", 2018).

The first official record of mohair in Europe was when a French botanist wrote in his book, "Travel to the Levant" (1654), that the best goats in the world were raised in Ankara. Tournefort reported in 1653 that "Angora goats dazzle with their whiteness and have fibres as fine as silk" (Hunter and Hunter, 2001). In Ankara, mohair spinning was originally done by women to support their families. Later, when the export of unprocessed mohair was banned by the Sultan, a closely guarded mohair industry developed in Turkey. In 1838, the ban was lifted under pressure from England and a few bales were sent to Europe (Hunter and Hunter, 2001). The rapidly rising demand for raw mohair caused Angora goat breeders to crossbreed, and even for a while, there was the danger of Angora goat extinction ("Mohair Association", 2020).

Commercial Mohair spinning started in England in 1853. When mohair first arrived in Europe, wig makers understood and appreciated the unique qualities of mohair. Mohair products were first produced in England in the nineteenth century, fabric containing a mohair weft and a cotton warp being in high demand in 1883 (Hunter and Hunter, 2001). After the rapid development of the mohair industry and the breakthroughs made in the weaving industry of England, which was the biggest customer of mohair yarn in the middle of the 18th century, and the introduction of synthetic fibres into the textile industry, Ankara became unable to sell its mohair fabric and its mohair industry declined rapidly, at which stage mohair yarn began to take the place of mohair fabric in its foreign trade for a while, followed by raw mohair trade after the 19th century. In this way, economic life was able to maintain its vitality for a while. However, towards the end of the 19th century, Ankara's monopoly in mohair production came to an end when Angora goats arrived and were bred in South Africa and America. Despite all the negative changes it went through, Ankara mohair production had an important place in the country's economy during the Republican Period of Turkey. During the first years of the Republic, Angora goat breeding and mohair production also improved and stabilised, the number of goats increasing by some 55%. Nevertheless, the general economic crisis that swept the world between 1926 and 1928 affected both Turkey and mohair production. In the 1930s, attempts were made to revive the mohair production and mohair weaving industry by improving the breeding conditions of Angora goats ("Mohair and Sof", 2018).

Angora goats, which were bred only in Central Anatolia until the 1840s, were taken to South Africa (1838) and America (1849), adapting well to conditions in these countries. This breed, which was unique to Anatolia, became known all over the world as the Angora goat ("Other Goat Families", 2004). Angora goats were first taken to South Africa in 1838. Colonel Henderson, who was in the Indian army at that time, took 12 bucks and 1 doe to Africa, it not being realised that the 12 male goats (bucks) had been sterilised by the Ottomans before the voyage. Nevertheless, it so happened that the female goat (doe) gave birth to a male kid goat on the ship during the voyage. As a result of their crossbreeding with domestic goats, the foundation of today's South African mohair industry was laid ("Mohair Report", 2010).

The spread of the Angora goat to the USA dates back to the middle of the 19th century. The then Ottoman Sultan Abdülmecit asked the US President Polk for an expert on cotton production and Dr. JB Davis was appointed to study Ottoman cotton production and to make the necessary recommendations for improvements. After completing his duty in 1849, on his way back to the United States, he took with him 2 bucks and 7 does (ewes), which were given as a gift by Sultan Mecit to the then American President Polk ("Mohair Association", 2020). From this small beginning, the USA became one of the world's two largest producers of mohair ("Mohair Story", 2004).

In South Africa, the fibre or hair (i.e., mohair) of Angora goats, which farmers named H. Wos and W. Hopley initially started to raise for meat, began to attract the attention of other farmers. In 1851, 1856, 1857, 1858, 1860 and 1866 lots of goats were imported from Anatolia to the Cape with permission from the emperor in order to create a pure Angora goat herd. A farmer named Thomson exported the first batch of mohair in bales to England in 1870. An increasing number of Angora goats were imported into South Africa from Turkey between 1867 and 1880. Although a large number of goats died as a result of disease in 1881, South Africa's mohair exports increased in parallel with production, 1882 being recorded as the year when South African mohair began to be accepted as equivalent to Turkish mohair by American and British companies. The Zwarte Ruggens Farmers' Association (ZRFA), which paved the way for the formation of the Mohair Growers Association, was established in 1883 (Öktem and Atav, 2007).

South Africa's rivalry with Turkey worried Anatolian farmers and traders. Realizing that each goat and goat export represented a decreasing income for them, they pressed for the prevention of exports, but without success. Despite the opposition of the breeder producers, it was stipulated by a Law No. 21 of 1899 that an export duty of approximately $30 would be charged for each goat exported from the Cape Colony. In 1899, South Africa's mohair production exceeded that of Turkey and 56.3% of global mohair production occurred in South Africa, 40.5% in Turkey and 3.2% in the USA. (Oktem and Atav, 2007). Angora goats were first exported to Australia in 1856, but the Australian mohair industry only began to spread in the 1970s. Angora goats were taken from Australia to New Zealand in the 1860s (Hunter and Hunter, 2001), and in the early 1980s, some 20,000 Angora goats were exported from Australia to New Zealand (Hunter, 1993). Angora goats were introduced into England in 1881 (Hunter, 2020).

2.1.2. World Production of Mohair Fibres

Until the beginning of the 19th century, Turkey was the only mohair producer in the world, but towards the end of the 19th century, mohair production began to become widespread in certain regions of South Africa and the United States, especially Texas and California (Cook, 2001). Today, the most important mohair producing countries are South Africa and the United States. In addition, Turkey, Argentina, Lesotho, Australia and New Zealand can be listed among other significant mohair producing countries (Hunter and Hunter, 2001). Table 3 and Figure 5, show the distribution of global mohair production between 1970 and 2020 by country (Anon, Mohair South Africa, 2022; Hunter and Hunter, 2001; Hunter, 2020; "Fibre Report", 2020; "Fibre Report", 2021; "Fibre Report", 2022).

From Table 3 and Figure 5, it can be seen that there has been a continuous and dramatic decrease in mohair production since it peaked in 1988. In 1965, 30 million kg of mohair (15 million kg in the USA, 8 million kg in Turkey and 6 million kg in South Africa) were marketed. The drastic decreases in price caused production to drop to 13.2 million kg in 1975 (US, Turkey and South Africa all just under 4 million kg each). When the price of mohair increased in 1979 again, this triggered a significant increase in production, which peaked in 1988, at 26 million kg (South Africa 12 million kg, USA 8 million kg and Turkey 3 million kg). After 1988, mohair production started to decrease once more, initially very sharply but then slowing down ("The Angora Goat & Mohair Journal", 2011). The main reason for the decrease was a decrease in the demand for, and therefore price of, mohair due to changes in fashion (Öktem and Atav, 2007). Mohair demand and prices are very sensitive to fashion changes.

Figure 6 illustrates the share of global mohair production according to country as at 2020.

The world mohair production in 2021 was some 4.59 million kg. As can be seen in Figure 6, South Africa, Lesotho and Turkey are in the three main producers with a production of 50.76%, 16.34% and 10.24%, respectively. These are followed by Argentina with 7.84%, USA with 5.45%, Australia with 1.74% and New Zealand with 0.44% of production. The share of other countries in total production is 7.19%.

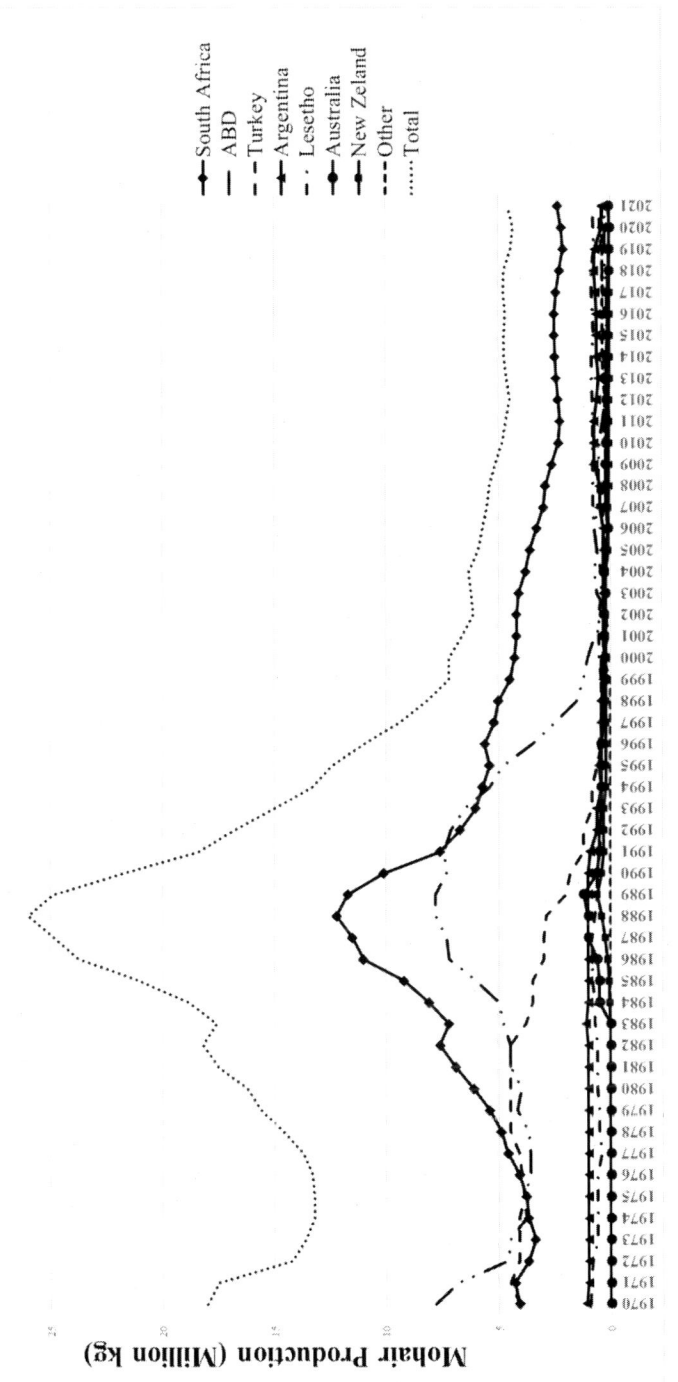

Figure 5. Distribution of angora production by country (drawn from data provided by Anon, Mohair South Africa, Hunter and Hunter, 2001; Hunter, 2020; "Fibre Report", 2020; "Fibre Report", 2021; "Fibre Report", 2022).

Table 3. Mohair production by country (million kg greasy) (Anon, Mohair South Africa, 2022; Hunter and Hunter, 2001; Hunter, 2020; "Fibre Report", 2020; "Fibre Report", 2021; "Fibre Report", 2022)

Year	S. Africa	USA	Turkey	Argentina	Lesotho	Australia	New Zealand	Other	Total
1970	4.1	7.8	4.1	1.1	0.9	0	0	0	18
1971	4.3	6.8	4.5	1	0.9	0	0	0	17.5
1972	3.7	4.6	4.1	1	0.8	0	0	0	14.2
1973	3.4	4.5	4.1	1	0.6	0	0	0	13.6
1974	3.7	3.8	4.1	1	0.6	0	0	0	13.2
1975	3.8	3.9	3.9	1	0.6	0	0	0	13.2
1976	4.1	3.6	4	1	0.6	0	0	0	13.3
1977	4.6	3.6	4.1	1	0.4	0	0	0	13.7
1978	4.9	3.7	4.5	1	0.5	0	0	0	14.6
1979	5.4	4.2	4.5	1	0.5	0	0	0	15.6
1980	6.1	4	4.5	1	0.6	0	0	0	16.2
1981	6.9	4.5	4.5	1	0.6	0	0	0	17.5
1982	7.6	4.5	4.5	1	0.6	0	0	0	18.2
1983	7.2	4.8	3.8	1.1	0.7	0	0	0	17.6
1984	8.1	5	3.5	1	0.7	0.5	0.05	0	18.85
1985	9.2	6	3.5	1	0.8	0.5	0.07	0	21.07
1986	11	7.2	3	1	0.8	0.6	0.14	0	23.74
1987	11.5	7.3	3	1	0.8	1	0.25	0	24.85
1988	12.2	7.8	2.9	1	0.7	1	0.4	0	26
1989	11.7	7.8	2	1	0.6	1.2	0.6	0	24.9
1990	10.1	7.3	1.8	1	0.6	0.6	0.4	0	21.8
1991	7.6	7.4	1.2	0.9	0.5	0.5	0.3	0	18.4
1992	6.7	7.1	1.2	0.6	0.4	0.5	0.3	0	16.8
1993	6	6.5	0.8	0.6	0.4	0.4	0.3	0	15
1994	5.7	5.4	0.8	0.4	0.4	0.4	0.2	0	13.3

Year	S. Africa	USA	Turkey	Argentina	Lesotho	Australia	New Zealand	Other	Total
1995	5.4	4.8	0.6	0.5	0.5	0.4	0.2	0	12.4
1996	5.6	3.5	0.4	0.4	0.5	0.4	0.2	0	11
1997	5.2	2.5	0.4	0.4	0.4	0.3	0.2	0	9.4
1998	5	1.5	0.4	0.4	0.4	0.3	0.2	0	8.2
1999	4.5	1.2	0.4	0.3	0.4	0.2	0.2	0	7.2
2000	4.3	1	0.4	0.25	0.5	0.25	0.2	0.3	7.2
2001	4.2	0.7	0.1	0.3	0.5	0.3	0.2	0.3	6.6
2002	4.2	0.1	0.2	0.3	0.5	0.3	0.2	0.3	6.1
2003	4.1	0.6	0.2	0.3	0.4	0.2	0.2	0.2	6.2
2004	3.8	0.7	0.3	0.3	0.4	0.3	0.2	0.3	6.3
2005	3.6	0.6	0.3	0.3	0.6	0.2	0.1	0.2	5.9
2006	3.3	0.7	0.3	0.3	0.7	0.1	0.1	0.2	5.7
2007	3	0.55	0.35	0.45	0.75	0.2	0.1	0.1	5.5
2008	2.9	0.5	0.35	0.45	0.75	0.2	0.05	0.2	5.4
2009	2.6	0.5	0.15	0.7	0.75	0.2	0.05	0.2	5.15
2010	2.3	0.48	0.17	0.7	0.75	0.18	0.05	0.2	4.83
2011	2.23	0.35	0.15	0.7	0.75	0.15	0.04	0.3	4.67
2012	2.32	0.21	0.19	0.6	0.77	0.16	0.05	0.2	4.5
2013	2.4	0.15	0.26	0.5	0.8	0.17	0.03	0.32	4.63
2014	2.45	0.15	0.2	0.6	0.8	0.17	0.02	0.35	4.74
2015	2.48	0.15	0.3	0.6	0.73	0.12	0.02	0.35	4.75
2016	2.48	0.15	0.22	0.6	0.76	0.12	0.02	0.35	4.7
2017	2.4	0.15	0.3	0.65	0.8	0.12	0.02	0.35	4.79
2018	2.24	0.23	0.34	0.71	0.8	0.06	0.03	0.35	4.76
2019	2.08	0.23	0.38	0.66	0.7	0.05	0.03	0.33	4.45
2020	2.16	0.23	0.46	0.36	0.74	0.01	0.03	0.33	4.32
2021	2.33	0.25	0.47	0.36	0.75	0.08	0.02	0.33	4.59

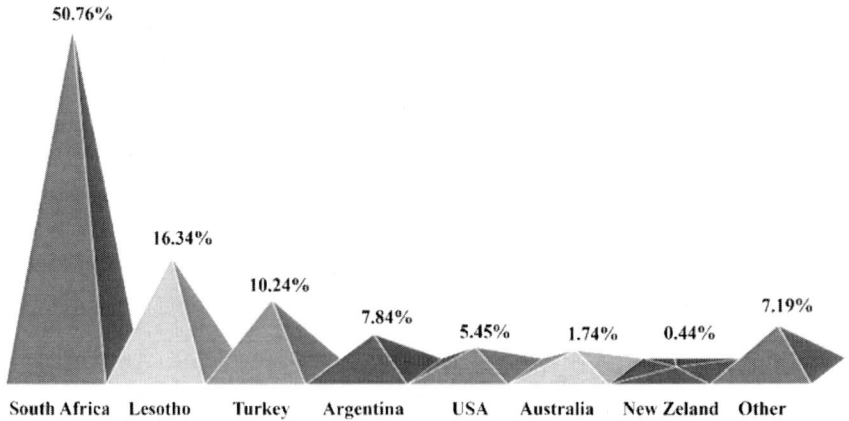

Figure 6. Share (%) of global mohair production according to country as at 2021 (drawn from data provided by "Fibre Report", 2022).

The world mohair industry originates in Turkey (Öktem and Atav, 2007), with the Angora goat being one of the most important gene resources of the country. Since the proclamation of the Republic of Turkey (1923), there has been a great improvement in Angora goats and the evaluation of mohair, and the export of breeding stock having largely been stopped (Vatansever, 2004). In 1930 the "Turkish Mohair Society" was established in order to improve the quality of mohair, and of Angora goats as well as to standardise the processing of the mohair and regulate mohair trading (Gürler, 2006). This society established the Turkish Mohair Society Sof Weaving House in Ankara in 1932 in order to revive the sof production and also started the weaving activity on the hand looms it provided. The Society also established an exemplary agriculture facility in Lalahan, near Ankara, in 1933. Goats from selected herd were given free of charge to the villagers or else at a very low price ("Mohair and Sof", 2018). The Turkish Mohair Society carried out important activities until 1951, when it was transferred to the Ministry of Agriculture, laying the groundwork for the establishment of the "Lalahan Livestock Central Research Institute" (Gürler, 2006). The "Turkey Wool and Mohair Incorporated Company (Türkiye Yapağı ve Tiftik A.Ş.)", a state-owned enterprise, was established in 1955 in order to increase mohair production in Turkey, officially starting its activities in 1956 (Şahin, 2013a). In 1959, more than half of the world's 11.3 million Angora goats were in Turkey ("Mohair and Sof", 2018). In 1969, the "Mohair Association (Tiftik Birlik)", which still continues its activities today, was established (Şahin, 2013b). The Association purchases mohair on behalf of the state in order to ensure price stability and protect the producers against foreign producers ("Mohair and Sof", 2018).

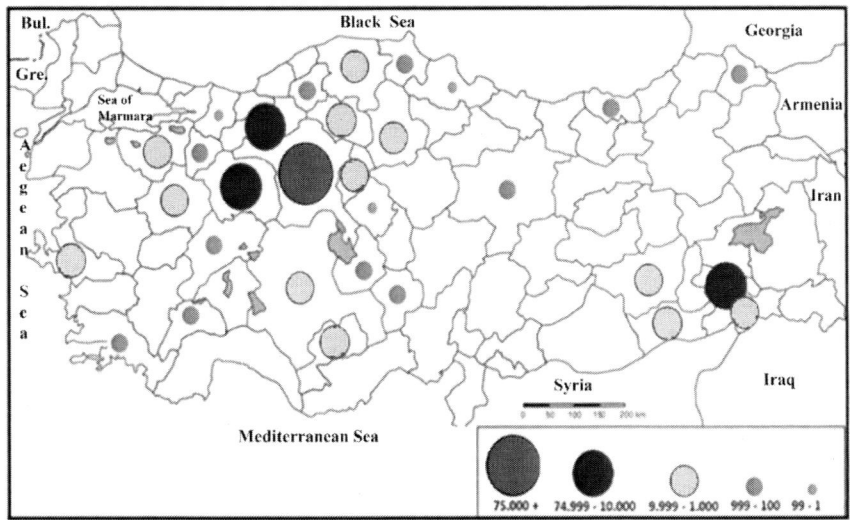

Figure 7. Geographical distribution of Angora goats in Turkey in 2010 (modified from Şahin, 2013a).

Angora goat ranks third in Turkey's goat population ("Mohair and Sof", 2018). According to 2010 data, the share of Angora Goats in the total goat population of Turkey is 2.4%, and its share in small ruminant population is only 0.5% (Şahin, 2013b). The purest specimens with the characteristics of the Angora goat are found in the Ankara region. Angora goats, previously occurred in all districts of Ankara are now concentrated in the northern districts of Güdül, Beypazarı, Ayaş and Nallıhan ("Mohair and Sof", 2018). Besides Ankara, Angora goats are found in Konya, Eskişehir, Çankırı, Afyon, Kastamonu, Kırıkkale, Yozgat, Çorum, Niğde, Bolu, Kırşehir, Karaman, Kütahya and Aksaray provinces of Central Anatolia and Siirt, Mardin, Şırnak and Batman provinces in Southern Anatolia (Öktem and Atav, 2007). Approximately 75% of Angora goats are located in Ankara ("Mohair Report", 2013; "Mohair Report", 2018). In Figure 7, the geographical distribution of Angora goats in Turkey in 2010 is given.

Due to the highly favourable climatic conditions prevailing in Turkey, the best mohair in the world was for a long time produced in Turkey. Turkish mohair is long, fine, very lustrous, distinctively white in colour and soft

(Öktem and Atav, 2007). The yield of scoured Turkish mohair is around 70-75%, with some 8% of the fibres coloured (Hunter, 1993).

As can be seen from Figure 5, Turkey, once the global leader in mohair production, it can no longer compete with South Africa in this respect. While South Africa was able to maintain a certain level of mohair production despite the changes in fashion, mohair production in Turkey has dropped enormously (Öktem and Atav, 2007), its annual production decreased from 3 million kg in 1988, to 0.421 million kg in 2000. Turkey, which was unrivalled in the world in terms of raw mohair, mohair yarn and fabric production and export until 1988, has lost its qualification as an important producing country today ("Mohair Report", 2017).

Mohair farming in Turkey continues as an inheritance from the father but has largely ceased to be an economic/commercial activity under present conditions. Nevertheless, the fact that all kinds of paid labour are done by family members and that some producers do not have other production opportunities, especially in mountainous regions, ensure that mohair production still continues today. In addition, another factor that prevents the extinction of the Angora Goat generation in Turkey is the "Premium and Direct Support Payments" practice ("Mohair Report", 2010). However, despite the implementation of this, decreased production continued after 2000 ("Mohair Report", 2017). It is calculated that Turkey's annual mohair production in 2010 was around 0.15 million kg (excluding coloured and black mohair production in the Southeast Region, since it is not an industrial product). Since this production is not sufficient for the domestic mohair industry's raw material requirements, Turkey became a mohair importer since the 1990s ("Mohair Report", 2010). Nevertheless, since 2012 the number of Angora goats and mohair production have begun to increase, the number of Angora goats increased from 158,000 in 2012, to 241,055 in 2019. Parallel to this, the mohair production increased from 0.2 million kg, 2012, to 0.379 million kg in 2019 ("Mohair Report", 2013; "Mohair Report", 2019).

Today's main mohair producer is South Africa, South African mohair popularly being referred to as "Cape mohair" (Hunter and Hunter, 2001). Mohair grown in Lesotho is called "Basutho mohair" (Hunter, 1993). South African Mohair is very fine and has a high yield, scoured yield being around 85%. The excellent quality of South African mohair makes it suitable for high-quality areas, such as the production of quality men's and women's clothing. South Africa exports some 95% of its production in a semi-processed or un-processed state (Hunter, 1993; Hunter and Hunter, 2001). In addition, South Africa buys and processes raw mohair from other producer countries and exports it as a semi processed material. Final value addition into products is generally carried out in European and Eastern countries ("Mohair and Sof", 2018). In South Africa, the South African Angora Growers Association was established in 1941, and the Mohair Advisory Board was established in 1951 (Hunter, 1993; Hunter and Hunter, 2001). Despite the strong organization in South Africa that covers not only the producers but the entire mohair industry as well as the existence of the "Mohair Trust", which acts as a kind of safety valve against price drops, and the modern and very effective marketing system, mohair production in South Africa continues to decline. Nevertheless, South Africa is relatively less affected by the overall decline due to it being so well organised, and still accounts for the World's most exports of mohair. Due to the excellent classing and quality of South African Mohair and extremely low levels or even absence of kemp as well as the possibility of shearing twice or at any time of the year, its mohair is highly rated and sought after ("Mohair Report", 2017). In 2020, South African mohair exports amounted to some 172 million US Dollars of which some 34% was to Italy, 31% to China, 11% to the UK, 7% to Taiwan, 4% to India, 42% to Bulgaria and 22% to Japan (Anon, Mohair South Africa, 2022).

Another important mohair producing country is the United States of America (USA). Mohair production, which is especially concentrated in Texas, is also carried out in New Mexico, Oklahoma, Michigan and other states, even if only in smaller quantities. Texas produces some 96% of US mohair, 90% of which being produced in Edwards Plateau, which has a mild dry climate and rough terrain making it especially suitable for raising Angora goats. The USA exports 98% of its production (Hunter and Hunter, 2001). The scouring yield of Texas mohair is around 73-82% (Hunter, 1993). In 1900, the American Angora Goat Breeders Association was established in Kansas City, Missouri, to further the Angora goat and mohair industry (Shelton, 1993). In 1966, the "American Mohair Council" was established as an organization that supports US mohair production, dealing with the marketing and development of mohair and also conducting research on mohair (Hunter, 1993). Angora goats bred in the USA have thicker bones, stronger joints and better early development ability than Angora goats bred in Turkey although their chest depth is lower. In Angora goats bred in Turkey, have a pelvis which has a wider pelvic cavity. Therefore, kidding is easier (İmeryüz, 1959).

New Zealand mohair is relatively fine but has a high kemp level, the latter also being the case for Argentine mohair, but the colour is good (Hunter and Hunter, 2001). The scoured yield of New Zealand mohair is around 88-90% (Hunter, 1993). There are approximately 550,000 Angora goats in the northwest of the Patagonia region in Argentina. Approximately 4,500 families make their living from mohair and the meat produced by Angora goats, most of the mohair being exported and only a small part is processed locally. Almost all Angora goats in Argentina are white ("Fibre Production", 2018). For Argentine mohair, the scoured yield is around 88-90% and the colour is good, but the kemp level is quite high as already mentioned (Hunter, 1993).

Australian mohair is quite fine and there are certain grades where kemp levels are low and even close to zero, whereas other grades have quite a high level of kemp. The scoured yield of Australian mohair is around 88-90% and the colour is good (Hunter, 1993). According to certain sources, there is a very small amount of mohair being produced in England with the mohair industry developing in this country (Hunter and Hunter, 2001). Angora goat were also imported and reared in other parts of the world: Madagascar, Mexico, China, Siam, Iran. Also, on the immense tableland of Puno in Peru (4.000-5.000 mts) flocks of Angora goats began to appear in the past (Alvigini, 1979). Apart from this, important initiatives have been made in recent years in the direction of mohair production in many European Union (EU) countries, especially Denmark, France, Italy, Spain and Portugal. The flocks are mainly based on imports from Australia, New Zealand and Texas with effective selection being applied in terms of mohair quantity and quality. However, since Angora goats are very sensitive to rain (humidity), systems should be developed in which two shearings are done per year in the rainy and cold northern EU countries. In such a system the goats have to be kept in shelters for a significant part of the year, which significantly increases the cost of mohair production. It is therefore recommended that mohair production in the EU should be in the dry and temperate regions of Southern Europe (Dellal et al., 2010). The "Angora Goat Breeders Association" was established in France in 1978, this Association importing Angora goats from Texas, and by pure breeding and mating Angora goat males (buck) and dairy goat does (ewes) there were 70 pure and 300 crossbred goats in France in 1983 (Vatansever, 2004).

Mohair has a wide range of textile applications, especially in clothing and home textiles, it is highly dependent on fashion as reflected in the large fluctuations in its price as illustrated in Figure 8 (Hunter and Hunter, 2001). The price of mohair in 2020 obtained for New Zealand Mohair Producers was as follows: $32/kg for premium quality Kids, $28/kg for Kids, $21.5/kg for Young Goats, and $17.5/kg for Adult Goats ("Farming Mohair", 2020). On the other hand, mohair prices in Turkey were set at 32 TL in 2018 by the Ministry of Agriculture and Forestry, and in 2019 this price was set at 52 TL for Kid Mohair and 44 TL for Adult Mohair ("Sheep and Goat", 2020).

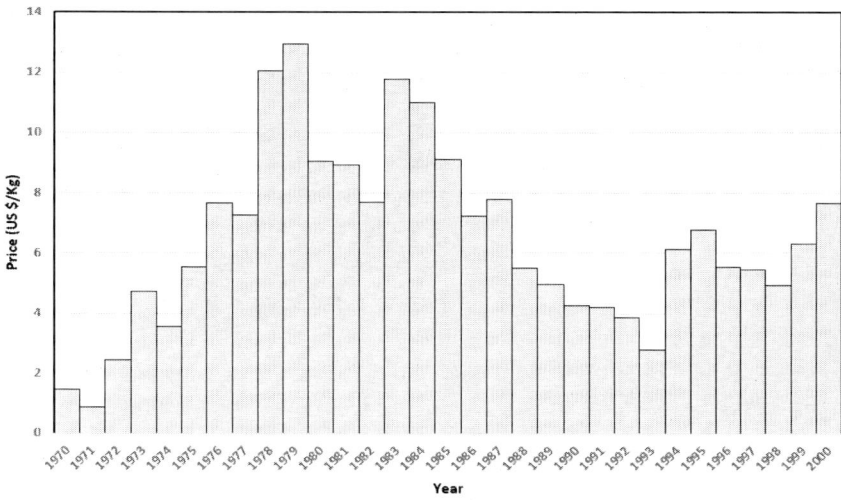

Figure 8. Fluctuations in South African mohair prices between 1970 and 2000 (reproduced from: Hunter and Hunter, 2001).

2.1.3. Harvesting Mohair Fibres and Factors Affecting Yield

Mohair fibres are obtained by shearing. Angora goats generally being shorn twice a year in South Africa, USA, Australia, Argentina and New Zealand (except when practical conditions (thorny vegetation, hooking etc.) demand a more frequent shearing, i.e., at four monthly intervals), and once a year in Turkey and Lesotho. Unlike cashmere goats, the Angora goat fibres grow continuously throughout the year and animals do not moult their fibres every year (Hunter and Hunter, 2001). However, spontaneous fibre shedding can occur, usually towards the end of winter, possibly triggered by shearing at the wrong time of year, although this is rare in well cared for Angora goats. In late winter, sometimes up to 25% of hair follicles (more in the case of older animals) can become inactive. As the fibres in the inactive follicles fall in the spring, the replacement fibres begin to grow which can cause matting (felting), the intertwining of the shedding fibres with the non-shedding fibres leading to the formation of matted areas. Fibre loss, due to adverse nutritional conditions may also occur as a result of some mineral deficiencies or imbalances, such as high calcium and low zinc (Hunter, 1993).

Mohair yield primarily depends on the genetic make-up of the animal ("Mohair Association", 2020). The heritability factor is 0.4, that is, the amount of greasy fibre is 40% dependent on genetic capacity (in another study 22% was found (Hunter and Hunter, 2001). Nevertheless, environmental factors, especially good care and nutrition also have an important effect. Mohair yield is low in under fed goats ("Mohair Association", 2020), related to the adult body mass of the goat (Hunter and Hunter, 2001).

The amount and type of fibre produced by an Angora goat depend on the number and ratio (S/P ratio) of follicles present in the skin, namely primary (P) and secondary (S) follicles. The Angora goat has almost the same skin follicle structure as sheep and its S/P ratio is between 7 and 12. It is generally thought that all primary follicles produce fibre when the kid Angora goat is born. Although the fibres that make up the hide are extremely coarse in newly born kids, the primary follicles later produce less coarse fibres. The fibres produced by the primary follicles in the kid at birth are called "mother wool/hair". These fibres are normally shed before the kid is four months old after which the primary follicles produce either normal mohair fibres or become dormant. Secondary follicles show little development during the first week after birth, whereas during the next two weeks, the maturation of the follicle is very rapid. By the time a well-nourished kid is six to eight weeks old, 75 to 80% of the final follicle count are producing fibres. Research has revealed that there is a significant relationship between nutrition and the number of follicles, with nutrition affecting fibre production. Nutrition is critical during late pregnancy (i.e., when the secondary follicles are developing in the foetus) and during the first ten months after birth (i.e., when secondary follicles mature and start producing fibre). If insufficient food is provided during these periods, the lifetime fibre production of the animal will be affected (Hunter, 1993). By adding sulphur to the feed of goats, significant increases are obtained in mohair yield as well as in fibre length. In addition, it is stated that adding iodine, copper, manganese, zinc and sulphur to the feed and a vitamin-mineral mixture increases live weight gain and mohair production, without affecting mohair quality (Vatansever, 2004).

Age (more correctly the associated weight and size of the goat) (McGregor, Butler and Ferguson 2012) also has a significant effect on mohair yield and fineness, less and finer mohair being produced by young goats than by the adult ones. The mohair yield starts to increase after the first shearing and this increase generally continues until the age of 4-5. After the age of 5 years, the mohair yield decreases ("Mohair Association", 2020). Young and Adult goats produce about 2 to 2.5 kg greasy fibre every 6 months. In Kids, the amount of fibre obtained is only 1 kg in the first shearing, and generally less than 2 kg in the second shearing (i.e., 1 year old). In addition, male goats usually produce significantly more and coarser fibres than females (Hunter and Hunter, 2001). Reproduction generally suppresses the growth rate of mohair (Hunter, 1993). The mohair yield of female goats decreases during the lactation period, the decrease being higher than during the pregnancy period (Hunter and Hunter, 2000).

Another parameter that affects the mohair yield is the season. More mohair is produced in summer than in winter. This is probably due to variation in day length and temperature as well as nutrition. Increasing day length and temperature increase the amount of fibre obtained, the amount of fibre produced by Angora goats can be 1.7 times higher in summer than in winter. Both in South Africa and in Texas the winter mohair tends to be slightly shorter, finer and less 'kempy' (Hunter and Hunter, 2001).

Utkanlar et al. (1964) investigated the effect of shearing twice per year on mohair yield, mohair quality and kid yield in Angora goats, finding that double shearing had an adverse effect on live weight, medullated fibre level, reproduction and birth rate; and a positive effect on fleece weight, ringlet length and fibre diameter.

One of the biggest factors affecting mohair yield is mohair fibre shedding, which can occur one or two months prior to shearing. It has been reported that in terms of mohair shedding apart from inflammatory diseases, external parasites and poisoning; relative humidity above 65% and temperatures above 10 degrees are highly effective especially in healthy animals (Arslan, 2005).

2.1.4. Classification of Mohair Fibres

In the mohair trade, the fibres are first shorn and then the fleeces are classed (Öktem and Atav, 2007). The simplest definition and sign of good classing is uniformity in length, fineness, style and character, and impurity (kemp, plant matter, and stain) content in each class. Therefore, an important goal of classing is to ensure that the quality, especially the fineness (diameter), is uniform. Classing standards and regulations are set and constantly updated in most of the major mohair producing countries. Therefore, parts that differ markedly in one or more important features within the fleece should be separated during classing (Hunter and Hunter, 2001).

Goats are age classified according to their age at the time of shearing and shorn separately accordingly. Fibre fineness is to a large extent determined by the age of the goat at the time of shearing, and it is therefore important to class the animals to be shorn into different age groups. Kids (six or twelve months), young goats (18 months) and mature goats (two years and older) must be shorn separately to simplify the classing of the mohair. Angoras may be shorn either by hand blade or by machine. Shearing by hand leaves more hair on the animal (10-20 mm) as compared to machine shearing. This may reduce the danger of losses during cold spells by providing some protection against the elements. When machine shearing is practised, the use of a comb-lifter designed to leave more hair on the animal could be considered. Goats are commonly crutched when they have approximately three months (75 mm) hair growth to avoid stained hair and twigs becoming entangled in the hair between the legs (Van der Westhuysen, Wentzel and Grobler, 1988).

All stained, lox and short pieces which may appear in the fleece, must be removed. The neck, which is usually coarser than the main fleece, is then removed, followed by the britches and any seedy hair which may be present. Should the fleece be very even (uniform), it can be kept intact. Often, however, the fleece has to be divided into fine and coarse (strong) parts, while some with good style-and-character should be kept separately. If the hair on the back appears to be very weathered, it should also be removed and packed separately. All bellies, britches, neck, ram and stained hair are kept and packed separately (Van der Westhuysen, Wentzel and Grobler, 1988).

The longest and coarsest fibres occur around the neck, especially in the lower part of the neck, and these fibres should be kept and classed separately. The fibres on the back and rump of the animal are generally shorter, finer, contain more kemp and foreign matter, and are more damaged. These should also be classed separately. The fibres at the shoulder, side, and thigh are generally intermediate in length and diameter, but are the best quality fibres in the fleece. The belly fibres can also be classified in this group if they do not contain stains or vegetable residues (Hunter, 1993).

As explained at the outset, the main object of classing is to class the various lines uniformly as possible. With this in mind, it is a good idea, once the shearing has been completed, to place the sorting table in front of the bin and to spread the hair from the bin thinly over the table. Any unwanted pieces can be detected and removed before the hair is baled, or bagged, and prepared for dispatch.

After classing, the fibres are packed and baled in special sacks or bales. Official brands and signs are stamped on one side of the bag or bale as is the case with wool bales, and special brands and numbers are stamped on the other side. Afterwards, each bag or bale is duly inspected by authorised experts (Öktem and Atav, 2007).

Mohair classing and grades vary in different countries. Below is information about the classification applied in the main producing countries, namely South Africa, United States of America and Turkey.

a) South Africa

In South Africa the classing is done by the producer at the time of shearing. In general, the best types of mohair are obtained from kids younger than 6 months old (first shearing) (i.e., Summer Kids). In South Africa, mohair from the first two shearings (i.e., 6 and 12 months) is usually defined as "Kid", that from the third shearing (i.e., 18 months) as "Young Goat", and the fourth and the fifth shearing (i.e., 24 and 30 months) and thereafter the mohair is classified as "Adult" (Hunter, 2020). The first shearing is classed as Summer Kids (SK) and the second shearing as Winter Kids (WK). For each age group different classing symbols are used for marking the hair as follows (Van der Westhuysen, Wentzel and Grobler, 1988):

- Kids (K)
- Young Goats (YG)
- Adult Goats (H)

A well-bred Angora goat will produce approximately 25 mm of hair per month and since Angora goats in South Africa are usually shorn twice a year (six monthly) the following length classes and symbols are used:

≥ 150 mm = A
125 -150 mm = B
100 – 125 mm = C
75 – 100 mm = D
50 – 75 mm = E

The lengths 150, 125, 100 and 75 mm represent the minimum lengths for A, B, C and D, respectively. Furthermore, the length variation within a class should not exceed 25 mm.

Mohair with good style and character is soft though firm to the touch, has the required amount of yolk (hardly visible under optimum conditions), is bright and lustrous and forms wavy, solid twisted staples, which have a uniform fibre length. Mohair is classed depending on its style and character as follows:

a. *Super and Good:* Super quality mohair has good style and character, has good colour, and is free from seed, stains and pigmented and kemp fibres. Such hair is indicated by the symbol S.
b. *Average:* Average quality mohair has reasonable style and character, tends to have a more straight and open staple formation, has a reasonable colour with a degree of kemp and is free from seed and pigmented fibres. No symbol.
c. *Poor:* Poor quality mohair lacks in style and character, is dull in appearance, has straight staples and a rough handle, which may contain some coarse kemp fibres as well as a degree of seed, but is free from pigmented fibres. No symbol.
d. *Mixed:* Mixed mohair consists of all badly bred, spongy, tender, matted and weathered hair and is indicated by the symbol M.

The short E-lengths cannot be marked as super for style and character.

General appearance also plays an important role in the classing of mohair. It is determined by the followings:

- Lustre (bright, not dull)
- Absence of foreign fibres (e.g., kemp, black, brown and other foreign fibres)
- Condition of hair (natural yolk, not greasy)
- Dust, stain and seed should be avoided
- An example of these symbols is given below for Kids (K):
  K-   Indicates hair shorn from kids.
  S-   Indicates above average style-and-character.
  F-   Indicates that the hair is fine (max. 27 μm).
  FNK-   Indicates all fine necks.
  NK-   Indicates necks, which are not over-strong (max. 30 μm).
  KSTN- Indicates lightly stained hair.
  KLOX- Indicates medium to heavily stained hair.

This first or second shearing of fine Kid hair of good style and character will be classed into an SFK – line, with the appropriate length symbol added e.g., BSFK. For the second shearing, the symbol K will be retained but less hair will qualify for the fine (i.e., F) symbol. A similar approach is applied to the other age groups, as appropriate.

- Seed contamination (VM): Seed contaminated mohair is separated into goat age grades and each into strong (coarse) and fine, and also into just two lengths viz. long and short. A and B lengths (i.e., ≥ 100 mm) are grouped as 'long seedy mohair' while D and E lengths (50 – 100 mm) are grouped as 'short seedy' mohair. Once seedy mohair has been sorted according to fineness and length it must be sorted according to the degree of contamination e.g., light, medium and heavy. For light contamination, attempts should be made to remove the seed by hand.
- Skin hair: Hair shorn from the skins of slaughtered animals must be packed separately and marked VEL, while hair shorn from animals which have died must be marked PLK.
- Marking ink or paint: Mohair contaminated by paint or other indelible marking materials must be offered for sale as BRANDS.
- Double cuts: Under no circumstances must double cuts be packed with the LOX, or any other lines. Provided it is clean and does not contain twigs or dust, it can be packed into a separate container and marked MOH. Required length – 20 mm and longer.
- Crossbred hair: Fleeces with excessive kemp, or hair shorn from crossbred goats, must be packed separately and marked with the prefix X. e.g., XFH, XH, XFYG, XFK, XK, XSTN and XLOX.
- Coloured hair: All hair containing grey, black, or brown fibres must be marked GREY (Van der Westhuysen, Wentzel and Grobler, 1988).

South African mohair is also classified according to the Bradford system which is based on fibre fineness as shown in Table 4 (Anon, Mohair South Africa, 2022).

Table 4. Classification of South African mohair (Anon, Mohair South Africa, 2022)

	Bradford (S)	Diameter Range (μm)
First kid	56'S-60'S	27-22
Second kid	50'S-56'S	30-28
Young goat	46'S-50'S	27-34
First adult	40'S-46'S	30-34
Second adult	36'S-40'S	35-40

b) United States of America

In the USA, breeders classify mohair according to the age of the animals and shearing seasons as follows (Harmancıoğlu, 1974):

a) Fall kid
b) Spring kid
c) 1.5 years old mohair
d) Adult mohair

The first of these refers to the mohair from the first shearing of six-month-old kids in autumn). The second refers to the mohair from the second shearing of one-year-old kids in the spring, the third refers to the mohair from the third shearing of 1.5-year-old animals, and the last refers to the mohair from the shearing of 2-year-old and older adult animals (Harmancıoğlu, 1974).

Moreover, according to the standards accepted by the American Society for Testing and Materials (ASTM) in the USA, a classification is made by considering the fineness and defects of the mohair. In this classification, basic quality is divided into four classes and the defective qualities into three classes (Harmancıoğlu, 1974) as follows:

Main Quality   *Defective Quality*
Kid
First   a) Stained (Urine stained, yellowish)
Second   b) Vegetable residues (Burry, filthy)
Third   c) Kempy (Dog hair)

Classification of dirty and clean American mohair can also be made according to the Bradford system and fineness, as shown in Table 5 (Kaymakçı, 2006).

Table 5. Classification of American mohair (Kaymakçı, 2006)

Commercial Classes	Dirty Mohair		Clean Mohair	
	Bradford (S)	Fineness (Micron)	Bradford (S)	Fineness (Micron)
Super kid	40'S<	23.0>	40'S<	23.55>
First kid	36-40'S	23.01-27.0	36-40'S	23.55-27.54
Second kid	30-32'S	27.01-31.0	30-32'S	27.55-31.54
First adult	26-28'S	31.01-35.0	26-28'S	31.55-35.54
Second adult	22-24'S	35.01-39.0	22-24'S	35.55-39.54
Third adult	18-20'S	39.01-43.0	18-20'S	39.55-43.54
Fourth adult	18'S>	43.01 and above	18'S>	43.55 and above

c) Turkey

According to whether they are subjected to the separation process or not, mohair fibres are divided into two groups as "standard" and "natural" mohair in Turkey. These groups are also divided into classes as "standard main mohair" and "standard secondary mohair", "natural main mohair" and "natural secondary mohair", and are also graded by considering the age, region and cleanliness of the Angora goat from which it is obtained. Classification of standard and natural mohair is given in Table 6 (Harmancıoğlu, 1974).

Table 6. Classification of standard and natural mohair (Harmancıoğlu, 1974)

Standard Mohair	
Standard Main Mohairs	*Standard Secondary Mohairs*
1) First kid	1) Coloured (taupe) mohair
2) Second kid	2) Greasy mohair
3) Fine mohair	3) Light mohair
4) Good mohair	4) Yellow mohair
5) Mediocre mohair	5) Skin mohair
6) Kastamonu mohair	6) Şekerî mohair
7) Konya mohair	7) Alâta mohair
8) Konya lowland mohair	8) Burry mohair
9) Gingerline mohair	9) Wizen mohair
Natural Mohair	
Natural Main Mohair	*Natural Secondary Mohair*
1) Kid (Natural)	1) Coloured (taupe) mohair (Natural)
2) Adult	2) Greasy mohair (Natural)
3) Fine mohair (Natural)	3) Skin mohair (Natural)
4) Good mohair (Natural)	4) Wizen mohair (Natural)
5) Mediocre mohair (Natural)	
6) Kastamonu mohair (Natural)	

7) Konya mohair
   a. Konya goods
   b. Oriental goods	

2.1.5. Microscopic Properties of Mohair Fibres

The cross-section, longitudinal view and the appearance of the scale layer of mohair fibres are shown in Figure 9.

From the longitudinal microscope images of mohair fibres, it is evident that they are quite even. The number of medullated fibres is generally not high, particularly in high quality mohair. A high number of medullated fibres (particulary kempy type) affects the quality of mohair. The cross-section of mohair fibres is slightly ellipted (oval) almost circular. Mohair does not have a very prominent scale structure. (Atav and Öktem, 2006).

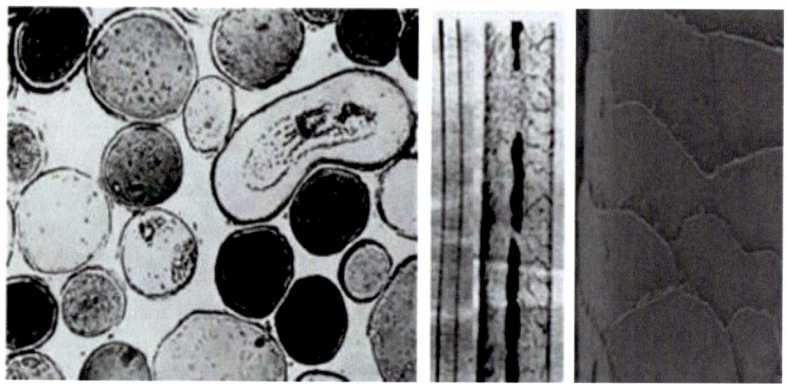

Figure 9. Mohair fibres; (a) cross section (Von Bergen and Krauss, 1942), (b) longitudinal view (fine and coarse fibre) (Yazıcıoğlu, 1996) and (c) view of the scale layer (surface structure) (X8000) (Atav, unpublished).

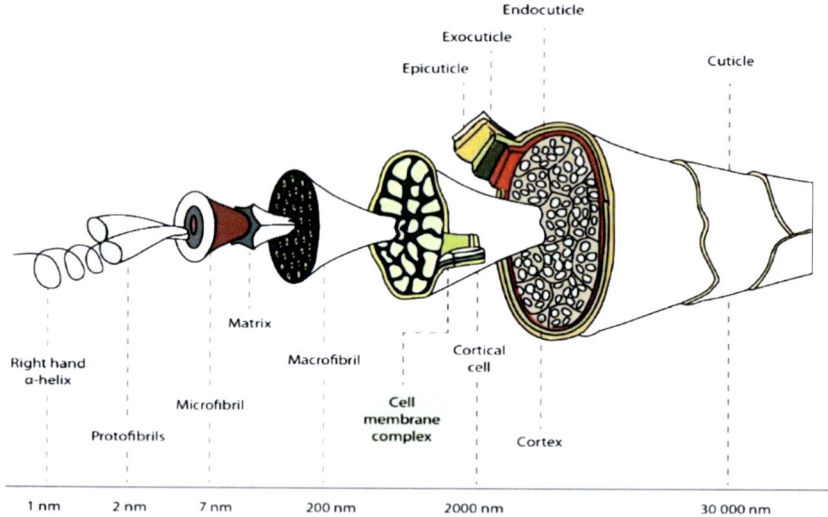

Figure 10. Mohair fibre structure (adapted from drawings by; Smith, 1988; Fraser MacRae and Rogers,1972; CSIRO, Australia).

As can be seen in Figure 10, mohair fibres generally consist of a cortex (cortical cells) layer (predominantly orthocortex) that forms the main and major part of the fibre, and epidermis (cuticle cells) layer, which consists of many overlapping scales. Sometimes there is also a continuous or discontinuous medulla (Hunter and Hunter, 2001). In addition, there are also kemp (objectionable medullated) hair usually with a very wide medulla and which are generally much coarser than the other fibres (Atav and Öktem, 2006).

a) Cuticle Layer

As in other animal fibres, the outer part of the mohair fibres is covered with a scale layer or epidermis (cuticle cells). The scales are thinner but wider than those of wool. The shapes of the scales tend to be different for fine, medium and coarse fibres (Harmancıoğlu, 1974). The scale layer forms a protective covering for the cortex and consists of three layers: epicuticle, exocuticle, and endocuticle (Figure 10). Each scale is covered by a thin semipermeable membrane, called the epicuticle, which contains protein and lipids. The cuticle layer is responsible for the slight felting behaviour of mohair fibres as well as the lustre of the mohair fibres. Although the mohair fibres have a similar appearance to wool fibres under the microscope, the scale layer of the mohair fibres is far less prominent and their upper edges not very raised (Hunter and Hunter, 2001). Hence, the angle they make with the fibre axis is also not as large as in wool fibres. The edges of the mohair fibre scales are not very prominent making the fibres brighter (more lustrous) and smoother (Harmancıoğlu, 1974).

Individual mohair scales have a larger surface than those of wool, the number of scales per unit length is less than in wool the case of scales, being about 5 scales per 100 microns for mohair, compared to 10-11 for fine wool fibres. The length of mohair fibre scales is therefore between 18-22 microns (Von Bergen and Krauss, 1942). The different scale structures of wool and mohair fibres allow the fibres to be distinguished from each other (Atav and Öktem, 2006). The average scale thickness (height) of mohair fibres is about 0.4 to 0.5 μm, while that of wool fibres, including lustrous wool (such as Buenos Aires), is about 0.8-1.0 μm. For wool fibres, this value rarely falls below 0.5 μm varying between 0.6 and 1.1 μm (Hunter and Hunter, 2001), illustrating the fact that the scale layer is not very prominent in mohair fibres, unlike in wool (Atav and Öktem, 2006). The scales are located much closer to the fibre body in mohair fibres (i.e., are flatter) than wool, resulting in mohair fibres being shiny (lustrous) and smooth. The lustre, smoothness, low friction, low felting tendency and some other properties of mohair are all related to its thin and large surface scale structure. In kemp hairs, the number of scales per

100 microns is about 10, which is twice that of normal mohair fibres (Hunter and Hunter, 2001).

b) Cortex Layer

The part of the mohair fibres under the cuticle layer is the cortex layer which forms the bulk of the fibre. This layer, as in wool fibres, consists of spindle shaped cortical cells arranged side by side (Harmancıoğlu, 1974), which in turn consist of macrofibrils (intermacrofibrillar matrix), microfibrils, protofibrils, and α-helices. The microfibril matrix complex largely determines the mechanical properties of the keratin fibre and also plays an important role in other physical properties. The microfibril matrix complex consists of partially helical, low-sulphur microfibrils embedded in a non-helical sulphur-rich matrix (Hunter and Hunter, 2001).

The cortex layer can manifest itself as distinct lines along the fibre length with many fibres having cigarette-shaped air-filled vacuoles of varying lengths. The proportions of fibres containing vacuoles in this way vary widely (Von Bergen and Krauss, 1942). As in wool fibres, the cortex of mohair fibres consists of two types of cells, called ortho- and para-cortex (Başer, 2002). The cystine (sulphur) content of the para-cortex is about twice that of the ortho-cortex, and hence the para-cortex is more stable than the ortho-cortex (Hunter and Hunter, 2001). The proportion of ortho-cortex cells in mohair is higher than that of the para-cortex (Başer, 2002). This together with the absence of a bilateral structure results in the "wavy", rather than "crimpy", structure of mohair fibre.

c) Medulla Layer

Some of the relatively coarse mohair fibres have an air-filled space, called the medulla, which occurs either in continuous, interrupted or fragmented forms, as in the case of wool. Although the continuous medulla is generally more common in mohair than in wool (Harmancıoğlu, 1974), multiple medulla being rarely encountered (Hunter and Hunter, 2001). In Figure 11, the types of medullae are shown (Wildman, 1954).

The problematic fibres, in which the medulla is very wide and which can be seen with the naked eye, are called "kemp" or more recently "objectionable medullated fibres" instead of "medullated fibres" (Hunter and Hunter, 2001). In medullated fibres, the diameter of the medulla is generally less than 60% of the fibre diameter, with that in "kemp", being mostly 60% or more.

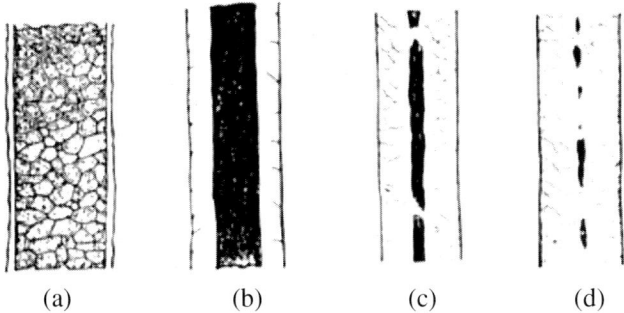

Figure 11. Types of medullae; (a) unbroken lattice (wide) (b) simple unbroken (c) interrupted (d) fragmented (Wildman, 1954).

All medullated fibres greater than about 100 μm in diameter are kemp. Mohair fibres with a diameter of less than 20 μm do not have a medulla (Lupton, Pfeiffer, and Blakeman, 1991). As with wool fibres, kemp can be distinguished from normal fibres by their white or opaque appearances, large medulla, and coarse appearance. These are fibres that are usually shed from the follicles and are described as dead, brittle and fragile. Towards the fibre end, their diameter decreases and the fibre becomes pointed (Harmancıoğlu, 1974). Kemp fibres are generally much coarser than the other fibres (1.8 times coarser on average) (Hunter and Hunter, 2001). Although the tenacity of kemp is generally lower than that of the mohair fibres, their elongation at break does not differ much (Hunter, 1993). In Figure 12, the cross-sectional and longitudinal views of heavily medullated (i.e., kemp) fibres as well as of the scale layer (surface structure) are shown.

Kemp generally has more than 10 scales in a length of 100 microns, namely twice as many as in normal mohair fibres and a possible means of distinguishing between kemp type and normal fibres microscopically (Harmancıoğlu, 1974). Kemp is usually quite flat and oval in cross section and are also mostly shorter than other mohair fibres. Most kemp, especially the shorter ones, can be removed during combing (and carding) and is reflected as processing loss (waste). Longer kemps are more difficult to remove by combing, and hence are highly undesirable. Kemp fibre diameter generally also increases with increasing mohair mean fibre diameter, as in the case for the medullated fibre content (%) in mohair (Hunter, 1993). McGregor, Butler, and Ferguson (2013a; 2013b) found that medullation increased with increasing mean fibre diameter and increasing animal body size.

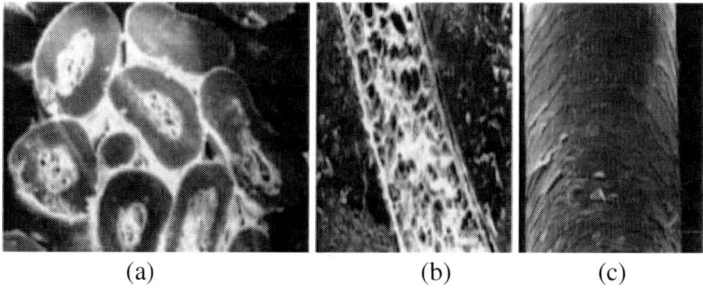

(a) (b) (c)

Figure 12. (a) Cross-sectional (b) longitudinal (c) the scale layer (surface structure) of heavily medullated (i.e., kemp) fibres (Hunter and Hunter, 2001).

Kemp fibres can be classified as short kemp, long kemp and heterotype fibres, short kemp usually being the most common. Fibres containing discontinuous (fragmented or discontinuous) medulla are often referred to as heterotypes. Heterotype fibres are therefore medullary (or "kemp-like") in certain sections along the fibre and normal in other sections. Heterotype fibres are generally longer and therefore more difficult to remove than short kemp fibres. It is stated that this type of fibres occurs more frequently in summer hair (Hunter, 1993).

The medulla generally contains cell residue and air pockets unlike cortical cells, medulla cells contain little or no sulphur. While the amount of medullated fibre normally does not exceed 1% in pure good quality mohair flocks, the proportion of medullated fibre may increase up to 3-5% with goat age and weight (Harmancıoğlu, 1974).

It has been found that kemp fibres have a different amino acid composition than normal mohair. The proteins of the medulla cells show a non-keratin structure and therefore exhibit a different chemical behaviour. They are easily broken down by proteolytic enzymes but show a high alkali resistance. Although the quantity of amino acids, such as glutamine, lysine and leucine, is higher in medulla cells compared to other cells, the amount of amino acids, such as glycine, serine, proline, threonine and especially cystine, is lower (Hunter and Hunter, 2001).

2.1.6. Physical Properties of Mohair Fibres

The processability, end-uses and general quality of mohair fibres in the textile field, and therefore also their value and price, are largely determined by the raw (greasy) mohair properties. Therefore, great effort has been devoted over the years, to the objective measurement of these properties to replace the traditionally used subjective techniques. Today, fibre diameter and yield, as well as many other properties, can be measured objectively with high precision. The main properties to be measured to characterise the greasy mohair fibres are:

- Fineness (diameter) and its variation
- Length and its variation
- Strength and its variation
- Elongation (elasticity)
- Clean fibre yield
- Medullation/kemp
- Vegetable and inorganic matter (such as dirt and stain) content
- Lustre
- Ondulation (waviness) and
- Colour

The most important parameter affecting the processing, price and end-use of mohair is fibre fineness. Kemp content and fibre length have a lesser, but still significant, effect on price and processability than the fibre diameter. Ondulation and lustre are considered important quality characteristics, not only because of their aesthetic appeal, but also because they may be related to fibres damage. Fibres that are shiny, have a pleasant handle, are uniform in fineness and length and do not contain kemp, contaminants (eg. foreign materials) or coloured/pigmented fibres are defined as quality mohair (Hunter, 1993).

Fineness (Diameter)

Fibre diameter (fineness) is usually the most important factor determining the quality and price of animal fibres such as mohair, the finer the fibre, the higher the demand and price. Fibre fineness determines processing behaviour and performance, as well as product type and quality. Fibre fineness determines the spinning fineness/limits and lightest fabric that can be produced from these fibres, as well as having a great effect on the handle of the fabric and the comfort of the garment produced from it (Hunter, 2020). While the average fibre diameter is the most important and parameter, diameter uniformity

(CV% of fibre diameter) is also of considerable importance in textiles (Hunter and Hunter, 2001).

Since the cross-sections of mohair fibres are very close to circular (round), as in wool, the fibre fineness can be determined by measuring the fibre diameter. Nevertheless, the ratio between the large diameter (D) and the small diameter (d) (ellipticity) is of some importance, as the cross section is not a complete circle and has some effect on spinning. Mohair fibres has been classified according to their spinnability as follows (Kaymakçı and Aşkın, 1997):

- D/d < 1,20 => very good spinnable mohair
- 1.20 < D/d < 1.22 => normal spinnable mohair
- D/d > 1.22 => weak spinnable mohair

The ratio between large and small diameters in mohair fibres is usually 1.12 or less (usually between 1.0 and 1.1 but rarely exceeding 1.2) (Hunter, 1993). Lower quality mohair fibres are generally less circular than good quality mohair fibres (Hunter, 1993). Coarse mohair fibres (kemp) with a very thick medulla have a distinctly elliptical cross-section (Shelton, 1993). Nevertheless, it is important to emphasise that the mean fibre diameter (fibre fineness), and ultimately the number of fibres in the yarn cross-section, has the major effect on spinnability and spinning limits (i.e., the finest yarn that can be spun).

The heritability factor for fibre fineness is 0.2 (in another study it was found to be 0.3), i.e., fibre fineness is 20% dependent on genetic capacity and is very sensitive to changes in diet and age and weight of the animal (Hunter and Hunter, 2001). Fineness of mohair fibres; differs according to age, body weight, gender and body parts (Kaymakçı, 2006). The age of the goat and the associated changes in body weight (McGregor, Butler and Ferguson, 2012) is probably the most important factor in determining the quantity and quality (fineness) of the mohair. At birth, kid goats largely have fibres that mainly originate from primary follicles. These also are the follicles that generally produce kemp and medullated fibres. From about 3 to 6 months after birth, fibres develop from the secondary follicles which produce finer and mostly medulla-free fibres, and the goats shed their birth hair (maternal hair) (Hunter, 2020).

As can be seen in Figure 13, fibre production increases after the birth of the goat and reaches a maximum between the ages of 3-4 years. With age, fibre diameter and also body mass increase, reaching a maximum at around 5 years (Hunter, 2020). The mohair fibres taper towards their tips since the fibres generally become coarser as the goat ages and grows heavier (up to about eight years of age) (Hunter, 1993).

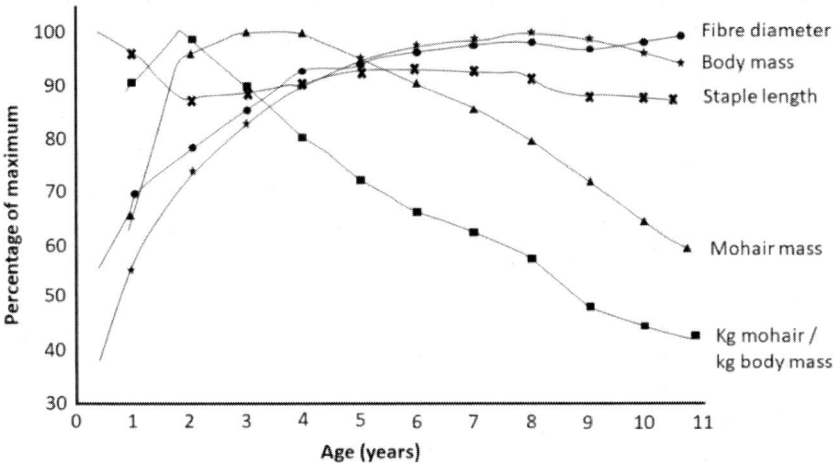

Figure 13. The effect of goat age and associated body weight on mohair yield and fibre properties (modified from Hunter and Hunter, 2001).

Fibre fineness in mohair occurs broadly speaking according to three classes viz, kid, young goat or adult depending on age (Harmancıoğlu, 1974). In general, Kid mohair is finer than about 30 μm (ranging from about 20 to 30 μm), Young Goat mohair is generally less than 34 μm (ranging from about 27 to 34 μm), and adult mohair is mostly coarser than about 34 μm (ranging between about 30 to 40 μm). The coefficient of variation (CV) of the fibre diameter is around 27% on average in mohair and can be considered largely independent of the average fibre diameter (Hunter, 2020). Within an age group the thinner the staple, and the softer the handle, the finer the fibre (hair), generally. Open, webbed fleeces of very soft handle usually contain the finest hair (Van der Westhuysen, Wentzel and Grobler, 1988).

While goats in Turkey are defined as young goats up to 3 years old, in South Africa they are considered as young goats up to 18 or 24 months old (Hunter, 1993). As age progresses, the quality of mohair decreases, the fibre becomes coarse and loses some of its elasticity and strength ("Mohair Association", 2020). If the increasing diameter of the fibres is not due to the aging of the goats, other causes should be considered such as increase in goat weight, lactation, nutrition, sudden climatic change. Disease can also be among these reasons, which can also cause changes in the fibre and decrease its commercial value (Harmancıoğlu, 1974).

Looking at the effect of gender on fibre fineness, it can be said that bucks generally produce significantly more and coarser fibre than does (Hunter and Hunter, 2001). Fibre diameter decreases during lactation. In addition, fibre diameter has been found to be associated with adult goat body mass/weight (Hunter, 1993). Castration is stated to have a positive effect on mohair production and fineness (Ankara) (Ertuğrul, 1991). Castrated bucks being called wethers. While bucks give 1.5-2 kg of mohair, wethers give 2-3 kg of mohair. Mohair from wethers is finer, less greasy and brighter than that from bucks. In large farms, 1/3 of the herd should consist of female goats (does), 1/3 of young goats and 1/3 of wethers and bucks ("Mohair Association", 2020).

Fibre diameter can vary significantly between fibres within the same fleece, and even within a staple (Hunter, 1993). When the variation of the fineness of the fibres according to the body regions is examined, it is noteworthy that the fibres in the neck and spine regions of the goat tend to be coarser than the other parts of the body (Hunter, 2020). In addition, the fibres produced during summer (shearing) are generally coarser than those during the winter shearing (Vatansever, 2004).

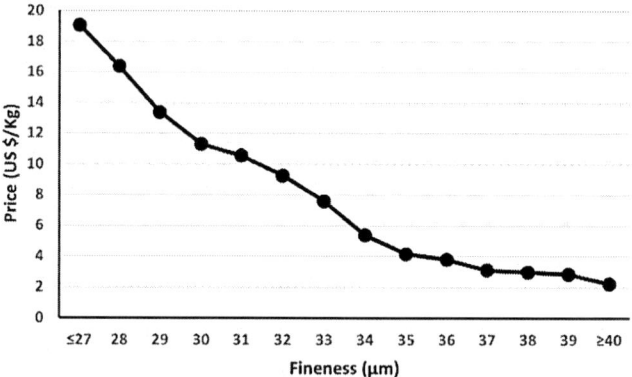

Figure 14. Price changes according to the fineness of South African mohair in 1999 (reproduced from Hunter and Hunter, 2001).

The fineness of mohair (mean fibre diameter) has a significant effect on its price and application. Figure 14 shows the variation in the price in 1999 according to the fineness for South African mohair. When the Figure 14 is examined, it is apparent that the change of even 1 μm in fineness can have a significant effect on prices, price of mohair fibres decreasing by some 5% with every 1 μm increase in fineness, up to about 34 μm after which the price does not change much with further increasing fineness (Hunter and Hunter, 2001).

Length

Fibre length refers to the length of fibre grown between two shearings ("Mohair Seminar", 1965). The heritability factor for fibre length is 0.8, hence fibre length is 80% dependent on genetic capacity. On the other hand, the length of the mohair shows little change with age (Hunter, 1993). The fibres show a length increase of about 20-25 mm per month (Hunter, 2020). Mohair is generally not shorn when it is shorter than 7.5 cm, that is, when it is less than about 4 months growth (Hunter, 1993). In a 6-month growth period, the fibre length can reach between 10-15 cm, while the fibre length can extend up to 20-30 cm after a one-year growth period (Harmancıoğlu, 1974). For this reason, when shearing once a year, the fibres have a length of around 30 cm, and about 15cm when shearing twice a year. The very long fibres (up to 30 cm) are commonly used to make women's braids, doll hair and theatrical wigs. The coefficient of variation (%CV) of fibre length is 40-70% in mohair and 20-25% in wool (Hunter, 1993).

Fibre length can differ from animal to animal, as well as varying according to the position on the body of the goat. The fibres are longest at shoulder and shorten from the front of the body to the back (Harmancıoğlu, 1974). In addition, the fibre length obtained in summer shearing is higher than that obtained in winter shearing (Vatansever, 2004). Although there is a significant relationship between fineness and length in wool fibres, this relationship is not as pronounced in mohair fibres as in wool. However, in general, it can be said that the shorter fibres are also finer, and the longer fibres are coarser (Harmancıoğlu, 1974).

The fact that the mohair fibres are undulated creates differences between their normal (or staple) lengths and their actual fibre lengths (Harmancıoğlu, 1974). For this reason, mohair length is divided into two as "ringlet (or staple) length" and "actual fibre length" as in wool. The length of the staple is the length of the mohair ringlet without being opened. The actual length is the length found after pulling the two ends of the mohair fibres and straightening the folds. Mohair can be classified according to the *staple length* as follows:

- *Short fibres:* those shorter than 8 cm
- *Medium fibres:* those between 9-12 cm
- *Long fibres:* those longer than 12 cm (Kaymakçı, 2006),

On the other hand, mohair fibres can be classified according to the actual *fibre length* as follows:

- *Short fibres:* those between 11-15 cm
- *Medium fibres:* those between 15-23 cm
- *Long fibres:* those longer than 23 cm (Başer, 2002).

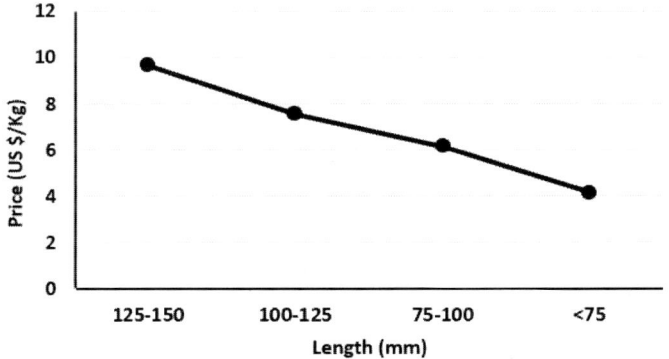

Figure 15. Price change according to the length of South African mohair in 1999 (reproduced from Hunter and Hunter, 2001).

In the production of mohair products, the fibre length is almost as important as fineness. Therefore, length plays an important role in the valuation of mohair (Kaymakçı, 2006). Fibre or staple length also has a significant effect on price, but this effect is less than that of fineness, the highest price is normally fetched by 15 cm mohair fibre, which represents about 6 months growth. Mohair below 7.5 cm has little economic value (Figure 15). For this reason, shearing more than twice a year does not provide any real benefit in terms of production efficiency (Hunter and Hunter, 2001).

Strength

Fibre strength is an important property in textile production. It has a significant effect on fibre breakage during mechanical processes, such as yarn and fabric production, as well as in determining fabric strength. In general, in the case of animal fibres, the force required to break the fibre, i.e., fibre strength, increases almost linearly with the fibre cross-sectional area (especially with the cross-sectional area of the thinnest (i.e., weakest) place along the fibre). On the other hand, the relative strength (tenacity) values obtained by dividing the fibre strength by the fibre cross-sectional area (preferably at the thinnest point) are almost constant for a given fibre type (Hunter and Hunter, 2001). In literature it is stated that mohair fibres have a greater yield stress than coarse sheep wool (36 S) and approximately the same initial Young's Modulus as the coarse sheep's wool (Meredith, 1945).

A study found that mohair had a higher relative strength (tenacity), modulus and elongation at break than wool of the same diameter. This is illustrated by the comparative values given in Table 7. In addition, lustrous wools (e.g., Lincoln and Buenos Aires) have a tenacity and initial modulus close to that of mohair (Smuts, Hunter and Van Rensburg, 1981).

Table 7. A comparison of the single fibre tensile properties of wool and mohair (Smuts, Hunter and Van Rensburg, 1981)

Feature	Wool	Mohair
Fibre diameter (μm)	18.1-33.1	20.7-44.3
Tenacity (cN/tex)	10.9-15.0	14.6-18.1
Initial modulus (cN/tex)	230-392	384-430
Elongation at break (%)	31.5-41.2	38.0-45.8

Experimental results obtained for fibre tenacity and elongation at break are related to the test length employed. Table 8 shows that as the test length increases fibre tenacity and elongation at break decrease (Smuts and Hunter, 1974).

In another study, it was found that bundle and single fibre tenacities were independent of fibre diameter (fineness), while the initial modulus increased slightly with an increase in fibre diameter (Hunter and Smuts, 1981). When the variation of the strength of the fibres according to the body regions is examined, it is seen that fibre strength tends to decrease, and fibre tenacity to increase as the fibres become finer from the shoulder region of the goat to the rump region (Vatansever, 2004). In one study it was found that mohair fibre strength changed with goat age. As according to Table 9 mohair fibre strength increases with increasing fibre diameter and age, while the fibre tenacity decreases (Davaslıgil, 1960).

Table 8. Tenacity and elongation at break versus test length for mohair fibres and kemp hairs (Smuts and Hunter, 1974)

Length (mm)	Tenacity (cN/tex)		Elongation at break (%)	
	Mohair	Kemp	Mohair	Kemp
10	17.9	14.8	47.9	46.6
40-50	15.5	12.4	38.3	34.9
100	12.6	8.8	30.2	21.8

Table 9. Changes in mohair strength and tenacity with goat age (Davaslıgil, 1960)

Feature	Age				
	1	2	3	4	5
Strength (g)	15.1	22.8	24.8	26.2	29.4
Tenacity (kg/mm^2)	30.8	29.3	25.6	23.7	24.7

Susich and Zagieboylo (1953) found that the wet strength of mohair is around 80% of its dry strength and is less affected by water than wool. The resistance of the fibres to swelling in water is dependent upon the morphology and chemical composition of the fibres.

Elongation (Elasticity)

The mohair fibre is very elastic and can return to its original shape even when it stretches up to 30% of its length. Due to the elasticity of the fibre, clothes made of mohair are resistant to wrinkling, bagging and shrinkage during wear (Hunter and Hunter, 2001). It is reported that the elasticity values of mohair fibres change depending on the age of the animal and this change is negative (Arslan, 2005). In their study, *Öztürk and Goncagül (1994)* determined the elongation (%) values for 1-, 2- and 3-year-old goats as 30.56±0.24, 27.39±0.20 and 26.81±0.19, respectively. In addition, elasticity decreases from the shoulder to the rump, and is higher for coarse fibres than for fine fibres (Vatansever, 2004). Moreover, in another study, fibre elasticity was found to be higher for summer fleeces than for winter fleeces (İmeryüz et al., 1969).

Bending Stiffness

It has been found that the static bending and extension modulus of mohair fibres were similar and of the order of 308 cN/tex. In addition, the medulla of kemp fibres differed in optical density, indicating different cell densities; this affected the bending but not the extension modulus. Two types of kemp, one with a filled medulla and the other with a virtually empty medulla, were investigated. For the latter, the bending and extension modulus of the kemp were similar at about 77 cN/tex, whereas the filled medulla gave a bending modulus of about 365 cN/tex, which was higher than that found for mohair (King, 1967). The extension modulus of the two types of kemp fibres were similar, indicating that any material in the medulla did not contribute to the tensile properties of the fibre, confirming the results of Hunter and Kruger (1966 and 1967).

Clean Fibre Yield

The fleece of the Angora goat, when shorn, contains natural and applied impurities, with the sweat (suint, the water-soluble component) and grease (wax) combined. Usually, 10% to 20% of the total fleece consists of material other than fibre. The grease is secreted by the sebaceous glands and the sweat by the sudiferous glands. Other natural impurities in mohair include sand and dust (i.e., inorganic matter), vegetable matter (e.g., burr, grass, seed) and moisture. Applied impurities include branding fluids, dipping compounds etc. (Hunter, 1993). Table 10 compares the moisture, grease and water-soluble content of raw (greasy) wool and mohair fleeces (Tucker et al., 1990).

The level and characteristics of the grease affect the quality (good or bad) of the mohair. The spread of the grease on the cuticle layer of the fibres ensures that the fibres are in close contact with each other without felting. If the amount of grease in mohair is less than normal, the protection of the ringlets against external factors will decrease, and thereby the important properties of the mohair fibres, such as colour, lustre and softness, and ultimately the value and price of the mohair will be adversely affected (Harmancıoğlu, 1974; "Mohair Seminar", 1965). Compared to wool, mohair fibres contain much less grease. For example, while merino wool may contain 15% grease, the amount of grease in mohair is around 4-6%. The grease content in mohair fibres obtained from kids and young goats is higher than that obtained from adult goats, with the grease content generally higher in winter than in summer. In addition, the amount of grease increases from the head to the posterior part of the animal (Hunter and Hunter, 2001) and from the tip to the root of the fibre (e.g., tip: 2%, middle: 4.6% and root: 6%). It is thought that an increase in feeding also increases the amount of grease in the fleece (Hunter, 1993). Since the amount of grease is less for mohair than wool, it is necessary to change the comb and scissors more frequently when shearing Angora Goats (Hunter and

Hunter, 2001), since grease prolongs the life of the comb and scissors due to its cooling and lubricating effects (Hunter, 1993).

Table 10. Moisture, grease and water-soluble matter content of raw wool and mohair (Tucker et al., 1990)

Fibre	Moisture (%)	Grease (%)	Water-soluble matter (%)
Wool	11.0-11.7	9.5-27.0	3.9-7.1
Mohair	12.0-14.4	1.2-8.0	1.8-4.2

Fibre yield is the expression of the clean mohair amount in % that will be obtained from a certain amount of dirty mohair under standard conditions after it has been scoured and cleaned of all foreign matter. As for sheep, internal factors from the animal's own organism (e.g., grease and suint) and external factors, such as dust, soil, fertiliser and vegetable matter added to the mohair fleece during the growth of the mohair, have an effect on the mohair yield (Atav and Öktem, 2006). Yield in mohair is very high, especially when compared to fine wool (Atav and Öktem, 2006), it generally being of the order of 80-90%, although it can sometimes be as low as 60%. The rest consists of grease, sweat and dirt (Hunter, 1993). The age of the animal does not have a significant effect on mohair yield (Atav and Öktem, 2006).

Characteristics, such as style and colour of the grease, play a major role in the classification and evaluation of Turkish mohair, since the grease affects the colour of the mohair and also its scouring process. The grease, which is not easily removed during scouring and remains on the fibres, reduces the value of mohair (Harmancıoğlu, 1974). The grease found in mohair can be classified as white, yellow, brown and reddish, depending on their colours. Among them:

- *White grease* is the most desirable because it generally can be easily removed during scouring and it is the colour of the white mohair.
- *Yellow grease* makes the colour of the mohair yellowish, but this is also considered acceptable as it can be scoured out easily.
- *Brown grease* makes the colour of the mohair dirty brown and is not easily removed during scouring. Therefore, it is not considered acceptable.
- *Reddish grease* makes the colour of the mohair red. It is not acceptable because it is sticky and difficult to remove during scouring (Harmancıoğlu, 1974).

Medullation/Kemp

The presence of kemp (objectionable medullated fibres) is generally the most undesirable quality characteristic of mohair (Hunter, 1993). Kemp in mohair is a source of problems in many areas, since the differ in appearance from the other mohair fibres. The main problems associated with the presence of kemp fibres in mohair are their chalky white appearance, their lighter (paler) appearance after dyeing and, to a lesser extent, their effect on the handle and prickliness of the fabric. The chalky white appearance of kemp fibres is mostly due to a reduction in the length of the light path in dyed fibres and the refraction of light within the fibre/medulla interface and the porous network structure of the cells within the medulla (aerian vesicles) (Hunter and Hunter, 2001). Studies have shown that the amount and rate of dye uptake for kemp fibres are the same as those for normal (i.e., solid) fibres, but they appear lighter in colour due to their different light reflecting properties (Hunter, 1993).

Although there are cases where kemp fibres are desired for special effects in certain types of carpets and woollen products, in general even a small amount of kemp in high-quality mohair has a marked negative impact on the fibre and product value and price. High quality mohair is largely free of kemp and medullated fibres, with kemp content well below 1%. Kemp levels can be controlled by selective breeding, but it may not be possible to completely eliminate it (Hunter, 1993). It has been reported that the kemp content of the mohair shorn during autumn is higher than that during the spring shearing (Yertürk, 1998). The kemp level varies between 2-6% in Turkish mohair, 1-3% in South African mohair and 2-4% in Texas mohair (Kaymakçı, 2006).

Vegetable and Inorganic Matter Content

Any undesirable contaminant, that will affect the quality of the final product or that needs to be removed, reduces the economic value of mohair. Similar to the presence of kemp, foreign matter in mohair, such as major burr and grass seeds, lead to a significant price reduction. Mohair containing undesirable vegetable matter can fetch on average half the price of other types of mohair. Therefore, burrs or excessive vegetable matter in the fleece should be removed. Urine and other permanent fibre stains affect the dyeing of the fibres, their value and the quality of the final product and precautions should be taken to reduce or remove these stains, especially urine stains (Hunter, 1993).

Lustre

Lustre is perhaps the most valuable property of the mohair fibre and is routinely evaluated subjectively during the marketing process. There is however no definitive objective method for determining the lustre of greasy fibres (Shelton, 1993). In one study, a goniophotometer was used to evaluate the lustre of mohair fibres. The goniophotometer provides accurate lustre information on single fibres, but the cost of the instrument and the cost of measuring many fibres per sample (to obtain statistically significant results for the sample) make this test method too expensive to be used in routine testing (Lupton and McColl, 2011). Although routine objective measurements can be made spectrophotometrically using a glossmeter today, it should be noted that precise results cannot be obtained for the lustre of textile materials on a routine basis.

Lustre remains one of the most sought-after properties of mohair fibres, and a lack of lustre in mohair often leads to a decrease in price (Hunter, 1993). Like colour, greasy mohair lustre is affected by the amount of grease and dirt present in the mohair. Therefore, it is necessary to evaluate the washed (clean) fibres in order to determine the lustre accurately (Shelton, 1993). The lustre of the mohair fibres makes the colours of the dyed fabrics appear bright, lustrous and attractive. The lustre of the fibres is largely related to the reflection of light from the surface of the fibres. Not only the arrangement of the surface scale, but also their size and the angle they make with the fibre axis affect the appearance of the fibres. Mohair fibres generally have a significant advantage over wool in terms of lustre (Harmancıoğlu, 1974). In general, the lustre of mohair is related to the less protruding (i.e., smoother) surface structure of the mohair fibre (Leeder, McGregor, and Steadman, 1998). Since the upper edges of the scales on the surface of mohair fibres are not very raised, the angle they make with the fibre axis is not as large as in the case of wool fibres, and the edges of the scales are not on top of each other. This makes the mohair fibres appear smoother and brighter (Harmancıoğlu, 1974). Studies have shown that the lustre of textile fibres, such as mohair, is related to the scale structure of the fibres, especially the scale thickness (height) (Maasdorp and Van Rensburg, 1983). In some studies, a negative correlation was found between mohair fibre lustre and diameter (Hunter, 1993). McGregor and Stapleton (2016) investigated the factors affecting the colour (especially whiteness and lustre) of commercial mohair sale lots in Australia. They found that with an increase in felting, kemp content and average fibre diameter (up to about 30 μm), a decrease in lustre and an increase in yellowness occurred. Van Rensburg and Maasdorp (1985) found that the lustre of mohair was decreased by solvent extraction, heating and steaming. The inclination angle of the scales with respect to the fibre axis decreased as the mean fibre diameter decreased, resulting in finer fibres being more lustrous than coarser fibres.

Ondulation (Waviness)

Essentially two types of locks (staples) are recognised viz "ringlet" (tight lock) and "non-ringlet" (mostly flat lock type), although there are basically three primary types of mohair fleeces based upon the formation of the lock, viz. the tight lock type (solid twisted ringlet), the flat lock type and the fluffy or open type. Angora breeders generally prefer a well-developed tight lock, or ringlet, although some prefer the flat lock which is also associated with a very desirable type of mohair. The tight lock type has ringlets throughout almost its entire length and is usually associated with the fineness of the fleece, while the flat lock type is usually wavy, has large crimps (waves), and absence of ringlets and forms a more "bulky" fleece. This type is usually associated with heavy and coarser fleeces, and a reasonably satisfactory quality of hair. The fluffy or open fleece type is usually objectionable on the farm since it is easily broken and torn out to a large extent by the brush. Flat lock type goats generally produce more greasy mohair but of a lower yield, than tight lock types and tends to be coarser. Ringlet type mohair is also thought to be associated with a more uniform staple length. The different lock types are not considered to be identifiable after scouring. The flat lock type tends to remain that, over the age of the goat (except as young kids), whereas the ringlet type is not always uniform or permanent. For example, while the tight ringlet type is seen in kids, it can turn into another ringlet type with age or take the form of another ringlet type towards the posterior portion of the body (Hunter, 1993).

Apart from length and fineness, the undulation or waviness (crimp) in the mohair fibre is important feature since it increases fibre cohesion during mechanical processing and spinning. The number of crimps (waves) of mohair fibres is lower than in wool particularly merino wool (Arslan, 2005). Among the mohair fibres, the more crimped ones are considered acceptable. As the ringlet (staple) length increases in mohair, the fibre length also increases. The shape and frequency of ringlets are largely hereditary which is important in terms of breeding (Harmancıoğlu, 1974). In addition, goat age also has an effect on the nature and structure of the mohair ringlet, fibre ringlet nature (character) decreasing with goat age ("Wool", 2011). Furthermore, a negative correlation has been found between ringlet frequency (wave) and fibre diameter (Hunter, 1993). Mohair fibres can be divided into three classes according to their crimp property as follows (Kaymakçı, 2006):

- *Low:* Less than 1 crimp per inch
- *Medium:* 1 to 1.25 crimps per inch
- *High:* More than 1.25 crimps per inch

Colour

Although mohair fibres are generally white or off-white, the mohair from some animals can be coloured (e.g., brown, black or reddish) (Hunter and Hunter, 2001) as illustrated in Figure 16 and Figure 17. However, such coloured (e.g., black or red) fibres can represent a problem in the finished product, especially when dyeing to light shades. Hence, the presence of coloured fibre reduces the value of mohair (Hunter, 1993). The white mohair is the most widespread and highly-valued. The origin of the coloured varities is rather vague, namely whether these varities are other mutations or the result of crossbreeding through the years (Alvigini, 1979). The colour of these fibres comes from the melanin pigments in the cortical cells that make up the cortex layer. Reddish brown mohair fibres containing colour pigments are produced in Turkey and are known as "Çengelli" in Turkey (Hatemi, 1988) and "Gingerline" in the rest of the world (Hunter and Hunter, 2001). Angora goats, with coloured, tawny and black fibres are very common in the Siirt, Mardin, Şırnak and Batman provinces in the South Eastern Anatolia Region of Turkey ("Mohair Report", 2018). The mohair obtained from coloured Angora goats is used only in local hand-woven and not elsewhere due to the difficulties experienced in dyeing (Arslan, 2005). Pigments in goats consist of two main types: eumelanin and pheomelanin, which can be present or absent in varying combinations (Coloured Mohair, 2023).

Eumelanin is responsible for black-blue grey-chocolate brown colours with most goats eumelanin only occurs in one shade, unless it is bleached by the sun or changed by some other environmental factor. Eumelanic areas of mohair routinely fade with age so that black kids become blue grey. On the other hand, chocolate brown usually fades to lighter brown. Brown eumelanin varies from very dark to very light in shade. Truly brown Angora goats are rare, as most Angoras have black eumelanin (Coloured Mohair, 2023).

Figure 16. Example of white and coloured mohair fibres (adapted from "Gingerline Image", 2011).

Pheomelanin is responsible for tan, cream and red colours. The pheomelanic tans are extremely variable, and unlike eumelanin, they frequently vary on an individual goat. Some goats have dark tan as well as pale cream pheomelanic areas. Pheomelanin can vary from very dark to very light in shade. At the light extreme it is almost white. At the dark extreme it can be confused with the browns of eumelanin. Generally, pheomelanic colours have a reddish tinge in contrast to eumelanin, which is usually a flatter brown, with little red in it (Coloured Mohair, 2023).

Figure 17. Coloured Angora goats (adapted from "Coloured Angora Goat", 2023).

The position of the eumelanic and pheomelanic areas on the goat determines the basic colour of the goat, and the classification of goat colour depends on the specific pattern of the pigmented areas. White regions on otherwise coloured goats are not pigmented, and pigment cells are usually completely absent in such white regions. This phenomenon is called "white spotting". The final colour of the goat is due to the interaction of eumelanin (black/brown), pheomelanin (red brown/tan/cream/white) and white spotting (white) (Coloured Mohair, 2023).

Other Properties
Mohair: in addition to the properties described so far, has a pleasant handle, absorbs moisture, has good resistance to heat and to soiling and has a low tendency to felt (Hunter and Hunter, 2001).

The specific gravity of mohair is 1.32 g/cm^3 (Cook, 2001), with that of kemp about half that of mohair depending upon the degree of medullation (i.e., "kempiness") (Hunter, 1993). Mohair fibres, like wool, can absorb moisture up to 30% of their weight without feeling damp or wet. However, the surface of the fibres is water repellent due to the layer of wax and oily substance on the fibre surface (Hunter and Hunter, 2001). The moisture exchange and heat-related properties of mohair are close to those of wool. However, the commercial moisture value (allowance) for mohair is 13%, which is relatively low compared to that of wool ("Mohair Story", 2004). The moisture-related properties of textile fibres are very important in terms of the role in the comfort of the fibres and their behaviour during wet processing and drying. The moisture absorption and other related properties of animal fibres, such as mohair, provide the desired wearer comfort. Temperature and humidity also play an important role in the visco-elastic properties of wool and mohair (Hunter and Hunter, 2001). Mohair fibres also have effective insulation properties. Mohair, as in the case of wool, also provides a "buffering action" by liberating heat when absorbing moisture due to its heat of sorption (wetting) properties, with the reverse occurring when it releases moisture, thereby enhancing overall comfort.

Mohair, like other animal fibres, has a relatively low flammability. When exposed to a naked flame, it burns at a low temperature and tends to shrink. The flame produces bead-like residues (ashes), but the fibre will stop burning almost as soon as it is taken away from the flame (Hunter and Hunter, 2001). Mohair has good sound insulation and heat resistant properties. Therefore, they are ideal for use in textiles in public places (theatres, hotel lobbies, offices, etc.) ("Mohair Story", 2004). Because of their smoothness and other properties, mohair fibres generally show good resistance to soiling, and dirt is also relatively easy to remove from them (Hunter and Hunter, 2001).

Another important feature of mohair is their low felting tendency. Like wool, mohair fibres have a lower friction when subjected to friction from root to tip (i.e., along or with the scales) compared to rubbing in the opposite direction (from tip to root i.e., against-scale). The low against-scale friction compared to wool, which is one of the distinguishing features of mohair fibres, can largely be attributed to the less prominent and very smooth scale structure of the fibres. It is this feature that gives the mohair its low felting property. Mohair therefore has a very small directional friction effect (DFE) due to the fact that the thin distal edges in the fibres are very easily deformed and at the same time the surface of the fibres is not so rough due to its less prominent scales and scale structure. The against-scale vs with-scale friction ratio of mohair fibres is 1.1. While for that of wool, is approximately 1.8. The "scaliness $((\mu_2-\mu_1) \times 100/\mu_1)$" measured in the dry state is approximately 60 for a fine merino wool, while it is approximately 5 for mohair. When measured wet, the relevant values are approximately 16 for mohair and 120 for merino wool (Hunter and Hunter, 2001) Therefore, mohair fibres do not felt as easily as wool (Corbman, 1983).

Handle, which is one of the very important properties for textile materials, is largely determined by the fibre fineness (mean fibre diameter) in mohair and other animals. With respect to fabric "prickle or itchiness" it is the relatively coarse fibres (i.e., coarse edge) which play a critical role, particularly the presence and level of fibres coarser than 30 μm. Although natural yolk improves the softness of the handle, the dipping of Angora goats (at least 3 days in summer and 7 to 10 days in winter) before shearing gives mohair a kinder handle and better lustre (Hunter, 1993), even if this may not persist through the commercial scouring operation.

2.1.7. Chemical Properties of Mohair Fibres

Mohair fibres are similar to wool fibres in terms of chemical structure, consisting of a protein - keratin structure. The elemental composition of mohair fibres is as follows:

50% Carbon
21% Oxygen
18% Nitrogen
7% Hydrogen

3% Sulfur

1% Ash (Mineral matter)

The sulphur content of mohair varies according to the conditions prevailing in the region where the Angora goats are raised (Harmancıoğlu, 1974). Table 11 shows the amino acid content of mohair (Hatemi, 1988).

Although the protein components of merino wool and mohair show some remarkable similarities, the general chemical, physical and morphological properties of these fibre types differ in various respects. Nevertheless, there are also apparently differences between kid mohair and adult mohair, with the amino acid composition of kid mohair found to be substantially similar to Merino and Lincoln wool. In a study it was found that the amino acid values of wool and mohair were not very different except for tyrosine, aspartic acid, serine and threonine. The most obvious chemical differences are the relatively low sulphur (and thus cystine) content and the relatively high aspartic acid content of mohair compared to that of wool (Hunter and Hunter, 2001).

Table 11. The amino acid content (% mol) of mohair fibres (various researchers)

Amino acids	Ward, Binkley, and Snell, 1955	Leon, 1972	Puchala et al., 2002
Alanine	4.26	4.03	5.62
Arginine	8.94	8.53	7.11
Aspartic acid	7.32	7.24	7.17
Cystine	-	9.70	11.67
Glycine	4.77	4.84	6.61
Glutamic acid	14.20	15.52	12.56
Histidine	0.90	1.09	0.94
Isoloysin	3.90	3,57	2.84
Lysine	3.07	3.26	2.82
Leucine	8.14	8.70	6.80
Methionine	0.52	-	1.08
Proline	5.64	6.41	7.37
Serine	6.05	7.83	10.53
Threonine	5.62	5.74	6.25
Tyrosine	2.39	3,51	2.91
Valine	6.12	7.76	5.16

The effects of chemicals, organic solvents, temperature, sunlight and microorganisms on mohair fibres are very similar to those on wool (Cook, 2001). However, the relatively high proportion of ortho-cortex in mohair fibres causes these fibres to be more sensitive to some chemicals than wool. Therefore, it should not be forgotten that temperature and treatment time play an important role in mohair fibres treated with chemical substances. More care should therefore be taken when subjecting mohair to wet treatments which involve chemical substances such as scouring, bleaching, carbonizing, dyeing, and finishing. On the other hand, the high proportion of ortho-cortex in mohair fibre ensures better dye-uptake (Harmancıoğlu, 1974).

In a study conducted by Hunter, Smuts and Barkhuysen (1977), mohair was treated in liquid ammonia and the effects of both fibre linear density and treatment time on fibre super-contraction, strength, initial modulus, friction and elongation at break were determined. On average, super-contraction, elongation at break, and friction were found to increase with increasing processing time, whereas fibre strength and initial modulus showed the opposite trends. It was noted that prolonged treatment in ammonia produced some crimp in the fibres, especially in the finer fibres, but reduced the lustre of the fibres and also caused some yellowing. Another study carried out by Roberts (1977) investigated the effects of processing conditions, such as dry heat, steam, aqueous treatments, involving different pHs and oxidative agents, on mohair fibre mass loss, yellowing and urea-bisulphite solubility. The results were compared with those obtained on Corriedale wool of a similar average fibre diameter (32 μm). It was found that the mass loss due to the wet processes was higher for mohair than for wool, whereas there was no significant difference in terms of yellowing. Changes in urea-bisulphite solubility indicated that mohair was less modified than the Corriedale wool under the milder conditions but more under more severe conditions (Roberts, 1977). These differences were ascribed to the relatively high ortho-cortex content in the mohair. The urea-bisulphite solubility of keratin fibres generally decreases when subjected to heat and alkali, but increases when subjected to acids and oxidizing agents. Yellowing in a weak alkaline solution is highly temperature dependent, increasing considerably at temperatures above 50 °C (Hunter, 1993).

Mohair swells by about 12% in diameter in methanol, ethanol, n-propanol, and dimethylformamide, but shrinks in isopropanol (Hunter, 1993). In a study conducted by Ahmad (1972), the rate of sorption of some organic solvents and reagents by mohair and their effects on its mechanical properties were investigated. It was found that the sorption values of Lincoln wool were slightly higher than those of mohair for all the reagents studied. The mohair fibres were also weakened when soaked in the various reagents (Ahmad, 1972). Like wool, mohair is affected by bacteria and will turn mouldy if stored under humid conditions, especially under hot and dark conditions. Compared to wool, mohair is more easily affected by bacteria and moulds (Hunter, 1993). If the mohair is exposed to sunlight for a long time on the goats back before shearing, the sulphurous compounds in the fibre will generally be damaged, and dyeing will be affected, and the fibre strength and flexibility will deteriorate (Harmancıoğlu, 1974).

Mohair generally exhibits good resistance to sunlight and is widely used in products, such as curtains and rugs, that are exposed to sunlight during use (Hunter, 1993). In a study conducted by Van Rensburg (1978), the degradation of woven and knitted mohair fabrics by sunlight and ultraviolet light was investigated and was found that the results obtained with ultraviolet light did not always match those obtained with sunlight. The best protection against sunlight was achieved by applying a polyacrylate pigment binder, UV absorber or certain dyes (Van Rensburg, 1978). Greater exposure to atmospheric conditions generally results in lower cystine (i.e., disulphide loss) and higher cysteine (i.e., increase in sulfhydryl content). As for wool, damaged (e.g., weathered) fibre tips take-up more dye and also faster than undamaged root parts (Hunter, 1993).

2.1.8. End-Uses of Mohair Fibres

Mohair is a valuable raw material of the textile industry because of its lustrous, elastic, durable, UV resistance, moisture-absorbing, low flammability, good dyeing and dirt-resistance properties ("Other Goat Families", 2004). The textile use of mohair dates back thousands of years, its fibre has found application in almost every textile area imaginable (Hunter, 2020). Countries that traditionally imported and used mohair fibres include England, Spain, Italy, Belgium and Japan. Recent developing markets for mohair (especially coarse mohair) include China, Taiwan, India and the former Soviet Union countries (Shelton, 2007).

The reasons mohair is more in demand and has a greater production volume compared to other luxury fibres are as follows:

- fashion demands and trends,
- the versatility and diverse applications (product range from clothing fabric to household textiles to upholstery),
- the strength and durability of the fibre,

- Mohair fabrics are lustrous, stain-proof, non-pilling and have good dyeability

Fabrics produced from mohair fibres are not only extremely durable, but also have attractive features, such as generally being light and thin and not prone to creasing. Besides all these features, mohair is also special because it can be easily blended with wool and other fibres and significantly improves the quality and image of the product even if it presents in low proportions in the blend ("Mohair Story", 2004).

One of the main advantages of mohair-containing fabrics is that, in the right fabric construction, they can keep the wearer warm in winter and cool in summer making it a popular choice especially in Japanese men's suiting fabrics ("Mohair Story", 2004). In 1992 it was estimated that between 70% and 80% of mohair menswear was consumed in Japan, where the fabrics were considered best suited for hot and humid summers (i.e., tropical conditions). In Japan, 30/70 mohair/wool blend fabrics are preferred in men's winter clothing (Hunter, 1993). Mohair is considered a cool fibre in fine worsted type low weight tropical fabrics. On the other hand, in brushed products such as shawls, scarves, rugs, sweaters and blankets, mohair provides warmth without weight largely as a result of the loosely constructed hairy nature of the woven and knitted fabrics (Hunter, 2020). Because mohair fibres are heat resistant and provide high sound insulation, they are ideal for use in textiles in public places (theatres, hotel lobbies, offices, etc.) ("Mohair Story", 2004). It is also effective when made into linings due to its good moisture absorption and drape characteristics (Hunter and Hunter, 2001). It is also particularly suitable for home textiles such as upholstery fabrics, curtains and rugs, due to its durability (soil resistance) and elasticity (Hunter, 2020). Pile fabrics produced from mohair are used in upholstery, carpets, furniture, curtains, sofa covers and bedspreads. Mohair velvets are also popular and generally consist of mohair as pile and cotton as ground fabric (Hunter, 1993).

Although mohair is extremely popular in many end-uses, it has some limitations, especially in the area of skin contact clothing, due to its relative coarseness compared to some other luxury fibres (Hunter, 2020). Mohair can be used as 100% pure depending on the application, as well as in blends with silk, linen, synthetic fibres and wool in different proportions, in the production of a wide range of fabrics (Öktem and Atav, 2007). Traditionally most of the world mohair production is processed mainly in England (40%), and also in France, Italy and Spain due to their high-capacity mohair processing facilities. Most of the processed mohair is sold to other countries as tops, yarn or fabric

("Mohair and Sof", 2018). Traditionally, 60% of mohair is converted into hand knitting yarn, 15% machine knitting yarn, 12% into women's fabrics and women's accessories, 8% into men's fabrics, 4% upholstery and draperies and 1% into industrial products (Kaymakçı, 2006). Angora Goat's age and the associated fibre fineness determine the end-use of mohair, and fine mohair produced by Kids and Young goats is used in more special areas such as clothing, while coarse mohair obtained from Adult animals is mostly used in the production of carpets and certain outerwear products ("Mohair and Sof", 2018). Whereas fine mohair produced by Kids and Young goats finds applications in machine knitting, that produced by Adult goats finds application in hand knitting yarns. The finest kid mohair is mainly used in socks and light weight clothing. Short and coarse mohair, on the other hand, has limited application and is generally used in cheaper products. The mohair combing wastes produced during yarn production are used in wool/mohair blended coats, as well as in the production of carpets, blankets, hats and socks (Hunter, 1993), often involving the woollen processing system.

The use of a very considerable amount of mohair in knitting makes mohair demand highly sensitive for fashion and the welfare levels of consumers in the buyer countries ("Mohair and Sof", 2018), some 80-90% of mohair consumption (especially adult mohair) being affected by fashion (Hunter, 2020). Mohair is generally used by blending it with wool. In Italy and Japan, approximately 20% of the mohair is blended with wool to increase the lustre, durability and image of the clothes ("Mohair and Sof", 2018). One of the most important organizations in the world that provided certification services for mohair products and worked for the protection of the mohair quality and image was the International Mohair Association (IMA), which was established in England in 1974, but ceased to exist in 2001. The Mohair Council of America, Mohair Producers Association of New Zealand-MOPANZ, South African Mohair Board (replaced by Mohair South Africa and the Mohair Trust) are other important organizations (Göktepe and Şahin, 2000).

The "Mohair Mark" was introduced by the IMA in 1976 and registered in 35 countries in 1999. Figure 18 (a) shows the "Mohair Mark" (Hunter and Hunter, 2001). Since the demise of the IMA its "Mohair Mark" has continued to be mainly used in only two countries, namely in Belgium by the Velour Weavers and in Japan. In South Africa, Mohair South Africa has introduced its own "Mohair Mark" (see Figure 18 (b))

Luxury Fibres Obtained from the Goats 71

(a)   (b)

Figure 18. (a) IMA Mohair Mark (Hunter and Hunter, 2001) and (b) South African Mohair Mark (Anon, Mohair South Africa, 2022).

Table 12. Rules determined by IMA for using the mohair trademark (label) (Hunter and Hunter, 2001)

Material	Minimum Mohair Content (%)	Evaluation
Knitting yarns and garments	70	Gold label
	40	Silver label
Women's fabrics, blankets, scarves etc.	70	Gold label
	25	Silver label
Menswear fabrics	50*	Gold label
	25	Silver label

(*or at least 30% kid mohair (fineness 32 µm or finer))

The South African Mohair Mark represents a commitment to excellence and is a symbol of authenticity and luxury and was developed to help consumers select products that contain only the highest quality natural fibre, there are 46 companies worldwide registered to use the mark presently (Anon, Mohair South Africa, 2022).

The rules determined by the IMA for using their mohair trademark (label) are given in Table 12 (Hunter and Hunter, 2001).

Furnishing velours with 100% mohair pile, irrespective of the backing, came in for promotion under the Gold Label system, the Silver Mark being allocated to goods with a minimum pile content of 70% mohair. Finished woollen goods, such as stoles, blankets, scarves etc, with a minimum mohair content of 70%, had a Silver rating, while ladies piece goods, for apparel manufacture, had to contain a minimum of 25% mohair to qualify for promotion. A minimum of 70% mohair was required for the IMA Gold Mark in hand knitting yarns, with at least 40% for a Silver Mark (Hunter, 1993).

In summary, Mohair is used with distinction in various, and widely different fields, such as men's and women's outerwear; upholstery; curtains; wall-hangings; blankets, carpets and rugs shawls, hats and scarves; interlinings, decorative purposes etc and remains a sought-after product in the textile industry ("Other Goat Families", 2004). The above-mentioned end-uses of mohair fibres are briefly discussed below.

a) Knitwear Industry

Mohair is widely used in knitwear, mostly to give it a particularly soft, lustrous and brushed ("woolly") look. Knitting accounts for up to 80% of the global mohair production, but this industry is very sensitive to seasonal fashion changes. Large quantities of mohair have been used in women's sweaters in the past for knitwear. Mohair is often used in blends with other fibres, such as wool, cotton, acrylic, nylon, silk, alpaca, angora rabbit wool, and polyester (usually multifibre blends). Initially, mohair yarns were not considered suitable for machine knitting due to the protruding fibres or the looped yarn structure hooking to the needles, but currently mohair yarns are also used in machine knitting (Hunter, 1993). Headgear, socks, scarves and gloves machine knitted from mohair are in great demand in the autumn-winter season (Öktem and Atav, 2007). Figure 19 shows a hand knitted mohair sock ("Ayaş Mohair Sock", 2008).

Mohair blended with wool, bamboo and other fibres has found an important application in highly comfortable socks, especially for active sportswear (e.g., cricket, mountaineering etc.) and medical applications (e.g., for diabetics) (Hunter, 2020). Kid or Adult mohair (worsted spun) has been used in men's socks, while 25% mohair/75% acrylic mohair socks have also been used for hiking, ankle socks, girl's knee socks and socks for farmers. Mohair in socks is stated to mitigate sweating and odours. Mohair socks are popular with people who wear boots continuously because of wet working conditions. Mohair/nylon (for reinforcing) socks, including cushion (plush) soles, are also popular, particularly for outdoor, leisure and active wear (i.e., hiking, mountaineering and sports). Mohair combines good moisture management with high insulation and exceptional durability and is less prone to trap odour forming bacteria (Hunter, 1993), particularly when used in blends with other "comfortable" fibres, such as wool, bamboo and nylon, the latter mainly for reinforcement.

Since much heat is lost from the head in active sports, caps that completely cover the head are important. For this purpose, berets made from mohair fibres are excellent, their important advantage being that they do not irritate the face ("Mohair Beret", 2004).

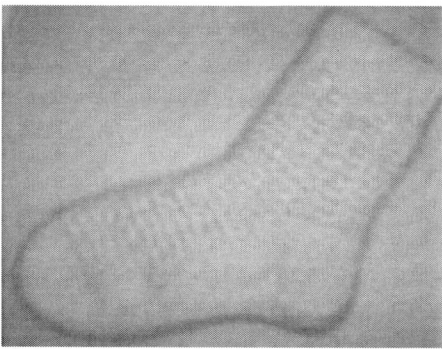

Figure 19. Hand knitted mohair sock ("Ayaş Mohair Sock", 2008).

b) Menswear Fabrics

Thin summer (tropical) suiting fabrics, fancy dresses, jackets and overcoats, sports shirts all fall under this category. The fabrics are typically woven with a wool or cotton warp and a mohair or mohair/wool weft. Mainly England, and also Italy, Japan, France and the USA are successful in the weaving and export of this type of fabrics ("Mohair Seminar", 1965). They are mainly classified as follows:

- *Summer fine fabrics:* A popular use of mohair (usually kid mohair) is in menswear fabrics, particulary summer and tropical (Panama) fabrics, often in blends with wool (rarely is only mohair used) but also in blends with other fibres. In these fabrics, mohair is generally used in the weft (often as singles yarn) with a two-ply fine wool as warp yarn. Relatively thin, light weight and smooth mohair containing tropical fabrics are famous for their coolness, good wrinkle recovery and good durability with mohair/wool/polyester blends having also been used for tropical clothing (Hunter, 1993). Although this type of fabric is classified as "summer", it can also be worn in winter. This type of fabric is called "Palm-Beach" or "Turkish-Mohair" ("Mohair Seminar", 1965).
- *Fancy dresses:* These types of fabrics include plain, Panama, serge etc. woven from yarns of various colours (uni, melange, etc.) produced by using long combing waste and noil and short mohair together with wool. The handle of these fabrics is crisp and these fabrics are suitable for winter clothes ("Mohair Seminar", 1965).

- *Elastic fabrics:* These fabrics, which are very difficult to produce, require great care during yarn production, weaving and finishing, and always require excellent raw materials. The purpose of using mohair in elastic fabrics is to achieve brightness and lustre ("Mohair Seminar", 1965).
- *Jackets and overcoats:* The use of mohair in this type of fabric gives the fabric a distinctive character in terms of handle and appearance. Another reason for using mohair in jackets and overcoats is that it keeps the fabric straight (undistorted) and thus the jacket or overcoat preserves its form after sewing. The main disadvantage of mohair in this type of product is that it can reduce abrasion resistance especially at the cuffs and buttonholes. For this reason, attention should be paid to the percentage of mohair and efforts should be made to determine the optimum percentages ("Mohair Seminar", 1965).
- *Sport shirtings:* In this type of applications, mohair is generally only used as an effect yarn because of its lustre. It is used as in a weft because of its proneness to abrasion and slippage in very thin shirtings ("Mohair Seminar", 1965).

c) Women's Clothing Fabrics

Mohair fibres have a wide area of use in women's clothing ("Mohair Seminar", 1965). In women's clothing (skirts, suits, jackets, dresses, etc.), blends containing kid and young goat mohair are used, and mohair is sometimes blended with wool to create two-tone dyeing effects (Hunter, 1993). The use of mohair in these fabrics is as follows:

- *In the form of pile or other fancy yarns:* Yarns in a wide variety of counts are used in these fabrics, from mohair rovings to fine yarns. It is related to the size of the curl and the appearance of the fabric. This type of coat fabrics is generally produced from a combination of worsted and woollen yarns. It is important that the mohair colours and curls are chosen well and harmoniously ("Mohair Seminar", 1965).
- *In fabrics to be raised:* Mohair is used alone or in a blend in all overcoat fabrics that are subjected to a raising process. Mohair is also used in the production of fabrics imitating animal fur i.e., in artificial fur ("Mohair Seminar", 1965). In the early 1870s, fur imitation products were produced using fluffy mohair fabrics (Hunter, 1993).

- *In fancy winter dresses:* They are similar to coats made with a pile and fancy yarns, their weight being less ("Mohair Seminar", 1965).

d) Shawls and Scarves
In shawls and scarves, the warp and weft are double-layered, but very slightly twisted and of a plain fabric construction of a low density. They are particularly piled. And even the cavities are filled with mohair and wool piles, thus keeping the wearer very warm. They are fluffy and very light ("Mohair Seminar", 1965).

e) Ties
Ties, once very popular in the USA and which are resistant to abrasion and wrinkling, were produced using very high-quality merino wool in the warp and mohair blended with 30-50% wool in the weft (Hunter, 1993), such ties also being produced by means of self-twist filament spun yarns in later years.

f) Lining Fabrics
Mohair was very popular in garment linings such as suits (sometimes in blends with wool, cotton or rayon) in the Victorian era (Hunter, 1993). In the past, the warp of lining fabrics was a wool and/or cotton yarn. The scarf was produced in mohair plain or serj constructions. In the past, lining fabrics were produced using wool and/or cotton in the warp and mohair in the weft in a plain or serge weave. However, as a result of the increasing use of synthetic fibres due to their lower cost, mohair has lost its place in the market ("Mohair Seminar", 1965).

g) Interlinings
In interlining products, especially coloured, coarse mohair is used together with hair and wool and is processed into yarn using the semi-worsted system ("Mohair Seminar", 1965).

h) Decorative Accessories
Mohair fibres are used as ornamental accessories in jackets, hats and shoes because they are lustrous and can be dyed into bright colours. Also, fibres up to 30 cm long are used in the production of women's hairpieces, doll's hair and wigs used in theaters ("Mohair Seminar", 1965). In addition, leather made from the skin of an Angora goat can be used for the manufacture of gloves,

purses, book bindings and novelties. Mohair skins are also used in the manufacture of teddy bears (Hunter, 1993).

i) Upholstery Fabrics

Mohair is used as a yarn in upholstery fabrics, as well as for decorative purposes as a separate yarn. In plush upholstery, pile and velvet fabrics are woven using mohair, as well as mohair blended with either cotton, wool or synthetic yarns, and dyed to the desired colour, for use in various upholstery fabrics ("Mohair Seminar", 1965). Upholstery mohair plush fabrics were popular in the 1890s, being used as automobile upholstery and as upholstery in railway carriages due to its durability, soil and stain resistance and other desirable characteristics. In 1924, all automobile upholstery in the USA was produced from mohair (Hunter, 1993).

j) Blankets

Mohair blankets constitute one of the traditional uses of the lustrous fibre. Among the characteristic features of mohair blankets, include their low weight, soft and silky handle, excellent insulation properties and luxurious appearance (Hunter, 1993). Mohair can be used, mostly in blends, in blankets which are defined as standard and luxury, or coarse ("soldier") blankets.

Figure 20. Siirt blanket produced from mohair fibre ("Siirt Blanket", 2020).

Mohair and wool blend blankets are mainly produced in a light weight, low density and raised form. These occur in various forms, such as plaid, striped or straight ("Mohair Seminar", 1965). In Turkey, mohair is used in the production of the Siirt blanket which is registered in the Siirt province. Siirt blanket is formed by weaving the yarns, which are obtained via hand spinning, in looms and feathering by shaping with steel wires. Siirt blankets are produced using a plain knitting technique. In addition to being used as a blanket, with the same technique and production in different sizes, it can also be used as a prayer rug or as a decorative product. The mohair fibres used in the Siirt blanket are not dyed, the original colours of the fibres from the goats being used (Figure 20) (Tunçel, Taşkaynatan and Erkuzu, 2020).

k) Carpets

Mohair is used in long or short pile carpets, rugs and mats (machine and handmade). Generally, medium quality mohair is used, except for handmade rugs, while kemp containing mohair is also used in certain carpets (Hunter, 1993). Until the First World War, mohair was used in the carpet industry with the manufactured carpets being exported to England and more especially to the USA. Mohair is generally used in a blend with wool in carpet yarns. It is recommended to use mohair in the manufacture of long pile, lustrous and "ostentatious" carpets, these carpets being in the luxury class since they are expensive. In addition, the skins of Angora goats are dyed to various colours and used in the form of hides instead of carpets, in homes and also sometimes by placing them on carpets ("Mohair Seminar", 1965).

2.2. Cashmere Fibres

Cashmere is a protein fibre obtained from Cashmere goats that live on the high and dry plateaus from northern China to Mongolia, including the Gobi Desert. These goats (Figure 21) have adapted their outer hairs to the climatic conditions in order to survive. Beneath these outer guard hairs is a much better-quality fibre (down), called "cashmere" ("Cashmere", 2003). Cashmere goats *(Capra hircus Laniger)*, also known as "shawl goats" or "Tibetan goats", have two fibrous coats (usually white): a coarse guard hair (outer coat) and a fine down hair (undercoat). The coarse guard hair provides a physical protection against rain, shrubs, thorns etc. (Hunter, 2020), while the fine down fibres enable these goats to withstand the extreme winter cold of the Central Asian highlands, their original habitat (Dalton and Franck, 2001). Generally, the fine down fibres grow during summer, preparing the animal to withstand the winter conditions. These fibres are then shed in the spring, as the animal will not need this highly insulating layer under warm, even hot, summer conditions (Hunter, 2020). The international abbreviation for cashmere fibre is WS ("Fabric Abbreviations", 2020), this referring solely to the fine down fibres and excludes the coarse guard hair.

Cashmere fibres are fine and soft like, making them one of the most sought-after animal fibres in the textile industry ("Heritage Cashmere", 2003). Cashmere even becomes softer and during wear, and these fibres are called the "Fibre of Kings" (Dalton and Franck, 2001). Cashmere is also called "Tibet hair", "soft gold" and "fibrous diamond". According to the US Wool Products Labelling Act, cashmere is defined as the "fine (dehaired) undercoat fibres produced by a cashmere goat, with an average fibre diameter not exceeding 19 µm, and with no more than 3% of the fibres (by weight) having an average diameter that exceeds 30 µm, and a CV of diameter not exceeding 24%". According to this definition, the coarse guard hair of the cashmere goat is not therefore considered to be cashmere (Hunter, 2020). Cashmere gets its name from the city of "Kashmir" in India, located in the western Himalayas, on the border with Pakistan. Very little fibre is produced in this region today; however, cashmere is grown mainly in northern China, Mongolia, Tibet and Afghanistan. Smaller amounts of cashmere are also produced in the Central Asian Republics, Iran, Australia and New Zealand (Dalton and Franck, 2001).

Cashmere fibre is also referred to as "Pashmina" in the literature. However, Pashmina is a special type of fine cashmere wool obtained from Changthangi (local name Changra) goats raised in the Ladakh region of India (Bhattacharya et al., 2004; "Pashmina", 2020) and is derived from the Persian word "pashm" meaning wool (Kaymakçı, 2006). Pashmina, the annual production of which is only around 50 tons, is also called "Indian cashmere" and is generally finer (10-14 µm), softer and warmer than cashmere (Ammayappan et al., 2011). In some sources, it is stated that products are marketed as Pashmina in order to add more value to the products obtained from cashmere (Weijer, 2011). The name 'Pashmina' seems to be used rather loosely, and it is sometimes difficult to separate myth from fact regarding the origins of the fibre. Pashmina has, on occasion, been claimed to come from the Ibex, but that is unlikely, and it is generally accepted to be fine quality Indian cashmere. Originally, Pashmina shawls and scarves were produced in Kashmir from hand-spun and woven fine cashmere fibres gathered from the ground and bushes where the goats had been feeding. India is not a significant producer of cashmere fibre but in Kashmir there are shawls of extreme fineness and softness which have been in the possession of various families for generations, and many such shawls are still produced locally. These articles are expensive, even by cashmere standards, and are now available on the European and American markets. Some, but not all, alleged Pashmina shawls on the market contain wool, and some, described as Pashmina, actually originate from areas other than Kashmir but are sold by suppliers who use the words Pashmina and cashmere synonymously. This has naturally led to a certain amount of confusion in the marketplace (Dalton and Franck, 2001).

Figure 21. Cashmere goat (adapted from "Cashmere", 2011).

Cashmere goats generally live in high altitudes with cold and harsh winters, their average lifespan being 7 years (Dalton and Franck, 2001). Although they are slightly smaller than Angora goats, they differ in size according to their origin and the regions where they live (Atav et al., 2003). The height of these goats is between 60-80 cm and their average weight is 60 kg for males and 40 kg for females, with the Cashmere goats in the Gobi Desert generally smaller (Dalton and Franck, 2001). These goats are strong, active and can with stand even the lowest temperatures (Alvigini, 1979). They are usually white in colour and have spiral horns (Hunter, 2020), the ones of the males are much more developed. The frontal-nasal profile is rectilinear. Both genders have beards (Alvigini, 1979).

2.2.1. Historical Background of Cashmere

The hair known by the name of "Cashmere" comes from goats reared especially in Central Asia: Sinkiang, Tibet, Gobi, China, Mongolia etc.

(Alvigini, 1979). The name "Cashmir" comes from the eighteenth-century English spelling of "Kashmir", which was part of the Raj under British imperial control. "Cashmere shawls" made from goat wool and other materials, were woven in Kashmir from the fifteenth and sixteenth centuries. Untreated goat wool was transported from Ladakh to Kashmir where it was woven into intricately patterned shawls (Weijer, 2011). Not all writers agree about the rearing of these goats in so-called Kashmir, that is the actual region in the North-West of India. On the contrary it seems that the hair and the down fibre of the cashmere goats were carried to this region (already at the time of the Great Mogul) to be woven, especially into famous shawls, with materials imported from Tibet and the Punjab, etc. It seems that all Cashmere goats, reared in various localities, derive from the *Kel* breed of Kashmir and from similar breeds: the Chaghu, the Gaddi of the Punjab, the Kaghani of Pakistan. In France the name *Cashmere* was given to the precious shawls made with the down fibre coming from Asiatic localities. According to the famous scholar H. Epstein among the ancestors of the cashmere goats there is the ancient "Falconeri goat" and probably also the "Hircus engagrus goat" from which it is likely that it inherited the under-coat (Alvigini, 1979).

Cashmere fibre has been known since Roman times. However, as the trade between east and west developed, it gained more importance. Cashmere yarn was introduced to the industrial world by the French inventor François Bernier. The famous explorer, who saw that cashmere yarn was used as a primitive clothing by the locals during his trip to the Himalayan Valleys in 1664, brought a few shawls to Europe, causing Europe to recognise cashmere for the first time (Akimoğlu, 2000). Cashmere was first recognised in Europe as a fabric woven in the Himalayas in the 17th century and used for shawls and scarves (Göktepe, Canipek, and Soysal, 2018). In the 18th century, English travellers commercialised this yarn in Europe, where a cold climate prevailed. During the same century, Asians (Persian and Mongolian empires) were already using cashmere and began sending cashmere shawls as gifts to the French Royal Family. Thereupon, the cashmere shawl entered the dowry chests of aristocratic families and became a fashion (Akimoğlu, 2000).

Although the origin of the passion for cashmere dates back to the 1600s, Napoleon and Josephine de Beau-Harnais had a great share in its spread. At the beginning of the 19th century, the softness and attractiveness of cashmere yarn began to be written about (Akimoğlu, 2000). It entered the palaces of Ranjit Singh, who conquered Kashmir in 1819, and Europeans turned their eyes to these Kashmir shawls for the first time. However, these shawls already had a long history in Central Asian trade before interest in these shawls and

European demand increased in the 19th century. Persian empires purchased cashmere shawls in what is now Iran at the beginning of the 16th century and distributed them as "robes of honour". Likewise, the Mongol empire initiated the tradition of gifting cashmere shawls to its allies. Mogul Emperor Akbar initiated shawl cloth production in imperial workshops at Lahore, Patna, and Agra (Maskiell, 2002). Josephine de Beau-Harnais, who was fond of cashmere shawls, created the fashion in the bourgeois society she lived in (Akimoğlu, 2000). Shawls were highly valued and seen as a clear sign of status and heritage items that women could have (Weijer, 2011). Josephine often visited the town of Louviers, which was known as the textile centre of France at that time, and was personally interested in the emperor's uniforms and clothes. It was in this small town that Josephine developed the idea of bringing cashmere yarn to Europe. For this purpose, Terneau, one of the first French industrialists, was sent to the Himalayas. On his return to France, Terneau proposed to bring to Europe valuable goats, a cashmere source living in the Himalayas. With the help of the government and the French Embassy in Istanbul, 3,500 goats were first transported to Istanbul on a gruelling journey, and then by ship to the city of Toulon. The French created today's cashmere phenomenon by improving the combing and care of the goats (Akimoğlu, 2000). A French company, founded by M. Audresset in 1836, led to the spinning of finer yarn towards the end of the 19th century and the more widespread recognition of cashmere in Europe as a fibre beyond shawls (Göktepe, Canipek, and Soysal, 2018). Audresset started weaving the yarn developed and industrialised in its own factory. For this reason, Queen Victoria awarded the Audresset family with a medal in 1845 (Akimoğlu, 2000).

In the early 1870s, fashion trends reversed, and cashmere shawls became obsolete, causing the market to collapse. The good times returned for cashmere around 1998, when major designers once again started featuring cashmere shawls in their collections. Cashmere was once again regarded as a unique and highly sought after luxury product which no self-respecting follower of fashion could do without (Weijer, 2011).

2.2.2. World Production of Cashmere Fibres

Most of the world cashmere production (over 90%) comes from China and Mongolia (Hunter, 2020). Iran together with Afghanistan, are the third largest cashmere producers in the world after China and Mongolia (Ansari-Renani, 2015). Other producers of cashmere include India, Nepal, Pakistan, Tibet,

Kazakhstan, Tajikistan and Kyrgyzstan. A small amount of cashmere is produced in these countries (Weijer, 2011). The *Cashmere* or *Kangra Goat* breed of Tibet lives in the Kangra valley, in the North-West of the Punjab, along the western border with Tibet (Alvigini, 1979). Apart from these countries, cashmere is also produced in Australia and New Zealand. Attempts have also been made to breed goats with fine down fibres in the USA (Colourado), South Africa (Kwazulu-Natal) and some other European countries (Spain, Italy, Norway) (Dalton and Franck, 2001).

In 1991 the average annual raw cashmere production worldwide was: China-10,000 tons, Mongolia-3,450 tons, Iran and Afghanistan 1,800 tons, Pakistan-600 tons, New Zealand-150 tons and Australia-65 tons. In 1999, China's production increased to 20,000 tons and Mongolia's production increased to 5,600 tons ("Historic Cashmere Markets", 2020). According to 2016 data, China produces 19,200 tons per year, followed by Mongolia with 8,900 tons ("Cashmere Wool", 2020). According to 2018 data, China's cashmere production decreased to 15,000 tons, constituting approximately 60% of global raw cashmere production. The 2018 cashmere production of Mongolia was 9,400 tons, while that of the Middle East countries were around 1,000 tons. According to 2018 data, the number of cashmere goats in China was 120 million, and around 27 million in Mongolia (The Schneider Group, 2020).

The bulk of cashmere in China is produced in the north-western provinces. The cashmere is obtained by plucking or combing the goats in the first few weeks of June, which is then packaged and shipped from production centres to export ports. Major buyers of such cashmere include European countries and Japan ("Heritage Cashmere", 2003). Chinese cashmere is the finest and best quality. The Chinese government's liberalization of the economy in the mid-1980s led to a somewhat chaotic period in which prices rose, quality fell, it became difficult and complex to obtain cashmere. This resulted in a decline of some 30% in knitwear sales according to Dawson International, a knitwear manufacturer and major fibre buyer. To restore order in both fibre distribution and quality standards and improve the quality of exported goods, the Chinese government introduced regulations in 1989 mandating testing and also established a Cashmere Foreign Trade Centre in 1990 to manage exports. This Centre organises four trade fairs each year to sell cashmere and sets limits on export prices. In 1991, the Chinese government passed another regulation, namely that all textile products from China require a label of origin, cannot circumvent quota restrictions, and can only be exported to countries that have signed bilateral trade agreements with China. However, with the general opening up of the Chinese trade, the cashmere market was no longer controlled and cashmere trading open to everyone. The physical distribution of cashmere is similar to that of camel fibres, but the difference is that many Chinese companies produce cashmere knitwear and distribute it globally. There is also a domestic market for cashmere knitwear in China (Dalton and Franck, 2001).

Although Mongolia is the world's leading cashmere supplier, exporter, it was only in the mid-1970s that cashmere began to be processed industrially in the country. The foundation for the industrial processing of raw cashmere was laid in 1976, with the establishment of a small pilot factory equipped with world-leading technology supplied by Japan with the help of the United Nations Industrial Development Organization (UNIDO). Later, with the support of the Japanese Government, the Gobi factory was built and started operating at full capacity in 1980 (Yondonsambuu and Altantsetseg, 2003). Cashmere produced in the central part of Mongolia is obtained by combing the goats once a year, as in China ("Heritage Cashmere", 2003). Mongolian cashmere is coarser than Chinese cashmere, but longer (Weijer, 2011). Combing is done towards the end of spring, when the weather is warmer. The cashmere is transferred to world markets through Russian agents. The USA has been the largest buyer of Mongolian cashmere products since 1945 ("Heritage Cashmere", 2003).

Afghanistan cashmere is generally in the same class as that of Iran and is coarser than Mongolian cashmere (Weijer, 2011). It is generally dark in colour ("Heritage Cashmere", 2003). Afghanistan exports around 1000 tons of cashmere on an annual basis, the main trading centre being Herat. Cashmere is harvested only in certain areas of Afghanistan, mostly in the Western provinces of Herat, Farah, Ghor and Badghis, and to a lesser degree in the surrounding provinces. It is estimated that only around 30% of the Afghan cashmere goats are currently being harvested. Afghanistan exports almost all of its cashmere in its raw (greasy) form. Most of the stages along the value chain occur outside of Afghanistan; in Afghanistan itself only the production and the harvesting take place. The only exception is an Afghan – Chinese Joint Venture that has recently been established in Herat, which involves a scouring line. Afghan cashmere was traditionally transported to Belgium through Iran, with Iranian traders playing an important role, which may explain why most of the cashmere production and trading are still taking place in the West. Verviers, in Belgium, used to be the main market centre for cashmere, as it was situated in the centre of the textile industry in Europe (Weijer, 2011).

Since Iran cashmere is coarser than that of China and Mongolia, its prices are generally lower ("Heritage Cashmere", 2003). Iranian cashmere is indeed coarser but also longer than cashmere from China and Mongolia (Ansari-Renani et al., 2013). Of the 25 million goats in Iran, 5 million are classified as cashmere producing goats (Raeini, Birjandi, Abadeh and Nadoushan) with the remaining goats producing only small quantities of cashmere. More than 90% of Iranian cashmere is produced in the eastern part of the country, mainly by two breeds of goat; Raeini in the Kerman and Birjandi in the South Khorasan provinces, respectively. The exact quantities of Iran cashmere produced and exported are not known but it has been estimated that the 5 million cashmere goats produce about 2,000 tons of cashmere annually, which is exported either raw underhaired (70%) or processed (30%). The major production, marketing and processing centres of Iranian cashmere are Kerman and Birjand; Kerman, Birjand and Mashad; Semnan and Mashad, respectively. Global trade of Iranian cashmere is accounted for by the four major processing countries, namely China, England, Belgium and Italy (Ansari-Renani et al., 2013).

Kyrgyzstan is one of the original places where cashmere goats were domesticated and is therefore an important genetic resource of cashmere goats. Indeed, some of the best cashmere in the world comes from a local breed of goat in Southern Kyrgyzstan. These goats are black, white or scarlet, some with black and white spots on their bodies. Since there is no incentive to produce higher quality fibre, many farmers have crossbred their local animals with Pridon and Angora goats. These hybrids produce more, but lower quality, fibre, that from the Pridon breed being dark in colour. They are less valuable in the international market as they are not considered to be genuine cashmere. The breed that produces ultrafine cashmere in Kyrgyzstan is known as "jaidari", meaning "local". As crossbred goats become more common in Kyrgyzstan, jaidari goats are in danger of extinction. This represents a loss of a valuable genetic resource, since very fine cashmere is rare (Kerven and Toigonbaev, 2010).

Australia's wild (feral) goat-based cashmere industry started in 1972. These goats were probably descendants of the original domestic goats from various European and Asian regions in the late 1800s (Hunter, 2020). Dawson International began supporting the development of the Australian cashmere industry in 1980 and conducted testing and product development with Australian cashmere. By 1989 Australia had produced almost 70 tons of raw cashmere fibre. However, in the mid-1990s, Australian cashmere prices declined and production fell (McGregor, 2002). Cultivated in both Australia and New Zealand, Australian cashmere is mostly white in colour and is generally longer, stronger, cleaner, softer, less crimped and more lustrous than that produced in other countries. The average fine down fibre yield is about 200 g per animal (Hunter, 2020).

Cashmere fibre production was started in Scotland in 1985 by a group of farmers to meet the demand of the local textile industry. Many farmers realised the profitability of cashmere production and formed the Scottish Cashmere Producers Association with the support of a government research agency and the textile industry. Compared to world cashmere production, Scottish cashmere production was quite small, there being only about 50 cashmere farmers and some 2,500 cashmere goats, with a target of at least 10,000 cashmere goats ("Cashmere Scotland", 2003).

Cashmere fibre production initiatives started in some European countries, initially in Scotland. After the establishment of the European Union (EU), due to the emphasis on fine animal fibre production within the union's agricultural policy, it also started in other countries, such as Spain, Italy, Portugal, England, Germany, Norway and Denmark. Contrary to wool and mohair, the quality characteristics of cashmere are less affected by non-genetic factors, especially nutrition, and this positively affects the development of production systems for this fibre in the EU, especially in regions with poor agricultural land and difficult climatic conditions. Nevertheless, cashmere production ventures still lag behind those of mohair due to an inability to establish quality herds (Dellal et al., 2010).

The Cashmere and Camel Hair Manufacturers Institute (CCMI) was established in 1984 in USA to protect the integrity of pure cashmere and camel hair products through education, information and industry collaboration ("Identity and Mission", 2020).

The relatively small number of animals in the world from which cashmere fibres are obtained and the low yield of cashmere results in high prices. For example, it is necessary to gather the cashmere of at least 30 goats for one coat. The time required to obtain the amount of cashmere for one sweater from one goat is 4 years. Clearly, it takes a great deal of time and effort to transform cashmere into luxury garments. Hence, it follows that cashmere fibres and the clothes made from them are expensive (Atav et al., 2003). According to 2001 data, the price of cashmere fibres varies between 100-130 $/kg, that of fine and white knitting cashmere being 120-130 $/kg and lower grades from 100 to 110 $/kg (Dalton and Franck, 2001). The change in dehaired cashmere prices between 1968 and 2006 is given in Figure 22. It is clear from the figure that the world prices for cashmere are highly volatile, fluctuating widely over the years (Weijer, 2011). Figure 23, gives the changes in the price of Mongolian cashmere between 2006 and 2016 ("World Bank", 2019).

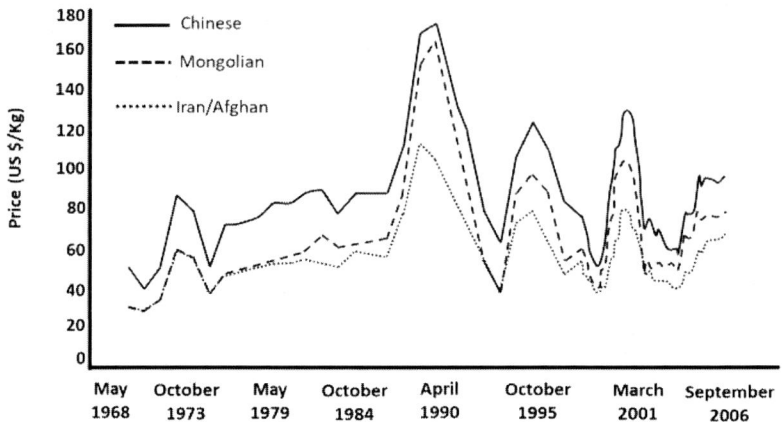

Figure 22. Prices of dehaired fine down (cashmere) fibres (reproduced from Weijer, 2011).

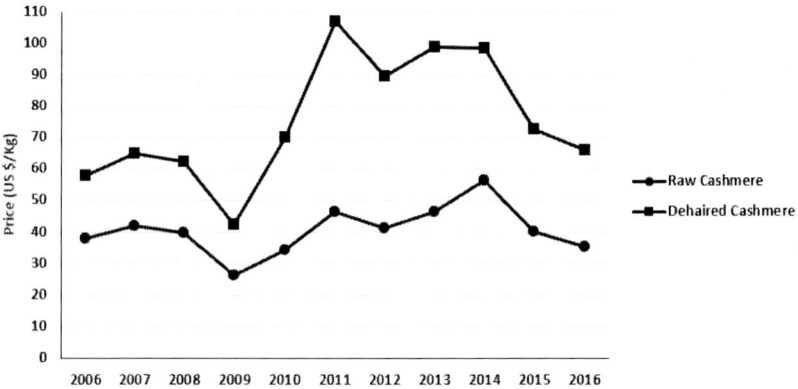

Figure 23. The change in the price of Mongolian cashmere between 2006 and 2016 (reproduced from "World Bank", 2019).

As can be seen from Fig. 22, Chinese cashmere price increased from 80 US$/kg in 1972 to 180 US$/kg in 1988 and decreased to 95 US$/kg in 2006 and once again increased to about 150 US$/kg in 2011, Mongolian and Iran cashmere prices also showing similar large fluctuations (Ansari-Renani et al., 2013). In contrast to this, prices for Afghan dehaired cashmere in 2006 were around 55-58 US$/kg for the European market (Weijer, 2011).

2.2.3. Harvesting Cashmere Fibres and Factors Affecting Yield

Cashmere and other related goat breeds essentially have two coats, namely a coarse outer coat (guard hair) and a fine down undercoat. The fibre (hair) of the two coats is completely different in terms of biological and textile properties (Dellal et al., 2010). The fleece of a cashmere goat grows from specialised follicles in the skin. The coarse hairs, that make up the coarse guard hair (outercoat), are produced only by the primary (P) follicles and they are characteristically medullated and coarse (>30 μm) and provides protection against the mechanical elements. Secondary (S) follicles are more numerous than primary follicles and produce non-medullated fine down fibres, called cashmere (<21 μm), which essentially provide thermal protection (Dellal et al., 2010; Ansari-Renani et al., 2013). The amount of cashmere (down) fibre produced by cashmere goats is significantly higher than that produced by the other relevant goat breeds (Dellal et al., 2010).

The amount and type of fibre produced by a cashmere goat depends upon the number of follicles present in the skin. The higher S/P ratio (secondary to primary follicles) and follicle density (the number of secondary follicles in one unit area of skin) the higher the cashmere production would be. It is generally thought that all primary follicles are producing fibre when the kid is born, the fibres that make up the birth coat being very coarse hair, whereas the secondary follicles produce the fine cashmere fibres. The secondary follicles hardly develop in the first week of the kid's life, but thereafter the secondary follicle development to maturity is very rapid. Research has demonstrated the close relationship between nutrition and follicle numbers and hence cashmere production. Since there are many times more fine cashmere producing secondary follicles than coarse hair producing primary follicles, it follows that the level of nutrition of the doe during late pregnancy (i.e., when the secondary follicles are developing in the foetus) and of the kid during its first ten months of life (i.e., when the secondary follicles are maturing and coming into production) are critical. If insufficient nutrition is provided at these critical stages, the lifetime cashmere production will be adversely affected (Ansari-Renani et al., 2013).

Cashmere fibres are harvested by either combing, shearing or collecting the moulted fibres. In China and Mongolia, cashmere fibres are harvested by combing during the 3 to 6-week spring season or by collecting the shed fibres from the ground, the goats shed their fibres in spring to protect themselves from the extreme summer heat. In Afghanistan, Iran, Australia and New Zealand, fibres are usually harvested by shearing (Dalton and Franck, 2001).

Secondary follicles shed their fine cashmere fibres (i.e., cashmere down) at the end of winter and beginning of spring this being called moulting. As a result of follicle inactivity, a sequential, bilateral-symmetrical pattern of cashmere shedding (moulting) initiates at the neck and proceeds in a wave posteriorly towards the rump, with up to a 5-6 weeks delay between the two sites neck and rump as illustrated in Figure 24 (Ansari-Renani et al., 2013).

Figure 24. Cashmere fibre shedding in Raeini goat; a sequential and bilateral symmetrical fibre shedding initiates at the neck (on the left) and proceeds in a wave towards the rump (on the right) (adapted from Ansari-Renani et al., 2013).

After the goats are 6 months old, a start can be made to harvest the fibre. As soon as the first fibres begin to be shed, the cashmere goat can be combed, and the entire herd must be combed every two weeks in order to obtain the maximum amount of cashmere fibre. Nevertheless, goats do not moult (shed) their fibres at the same time. Various types of combs are used for combing cashmere. It takes approximately 20-30 minutes to comb a goat (Süpüren Mengüç and Özdil, 2014). Antonini et al. (2016) state that combing all the goats at the same time by breeders and collecting cashmere from animals of different ages in the same bag reduce the value of the cashmere. They emphasised that combing the young goats first and then the old ones and keeping the fibres harvested from the upper parts of the body separate from those harvested from the other parts would be beneficial in obtaining batches with less contamination, since cleaner and finer fibres will be combed first and fleece parts containing more vegetable matter and impurities, such as faeces, will be kept separate.

In order to harvest the maximum weight of cashmere, the optimal time for a single shearing of cashmere goats would be at the end of the winter season, before follicle inactivity becomes substantial or before onset of shedding. At this stage goats are in their poorest body condition due to the cold weather and very limited feed availability. Thus, it is important from the point view of the welfare of the goats that some hair is left on the animal after harvesting the cashmere as this hair provides an essential protective layer against adverse weather conditions (Ansari-Renani et al., 2013).

The method of harvesting affects the fleece characteristics, such as the ratio of down fibre (cashmere) to guard hair, and the level of contaminants. Fleece characteristics are also influenced by nutrition and genetics (Hunter, 2020). Better nutrition results in longer and slightly coarser cashmere, but with significantly lower crimp (McGregor, 2004). It is stated that combing is more advantageous because the fibres obtained are cleaner, higher yielding (fine down fibre/coarse hair ratio) and longer, cashmere fibres being shortened when cut during shearing (Weijer, 2011). Commercial cashmere yields are higher from combed fleeces than shorn fleeces, making combed cashmere more attractive to textile processors (Ansari-Renani et al., 2013). Internationally, fibres with a length of at least 4 cm (40 mm) are in demand. Since it is not possible to cut the fibre against the skin during the shearing process, the fibre length is reduced by 1 cm (10 mm) or more. This reduces the market value of cashmere harvested by shearing (Kerven and Toigonbaev, 2010). Shearing and dehairing by hand can be seen from Figure 25 (Weijer, 2011).

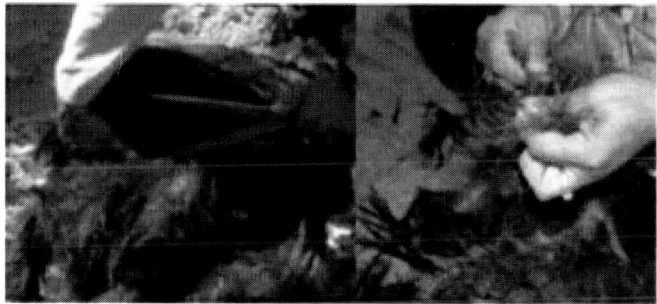

Figure 25. Shearing (on the left) and dehairing by hand (on the right) (adapted from Weijer, 2011).

Typically, cashmere processing commences with sorting, willowing (to remove dirt, grass, etc.) and aqueous scouring (to remove natural oils (grease), etc.) (Hunter, 2020). Cashmere sorting is the first process which greasy cashmere undergoes after it is purchased by the manufacturer. In the factory cashmere is sorted by hand according to fineness, length, soundness, colour (white, grey and brown) and level of vegetable matter, enabling the manufacturer to produce the style and quality of yarn or fabric for which the cashmere is best suited (Dalton and Franck, 2001; Ansari-Renani et al., 2013), coloured raw cashmere requiring more processing stages than white cashmere (McGregor, 2012). This is done quickly and requires considerable expertise. After sorting, the different categories (groups) of hair are 'willowed'. This is done by placing the fibres in a simple revolving machine to shake out most of the dust and grit. After sorting and willowing, the fibres are scoured before being dehaired (Dalton and Franck, 2001), the scouring yield being about 70% for cashmere. The residual grease, after scouring, should be around 0.5%. The most important step in cashmere processing is dehairing, the effectiveness of which has a major impact on the quality, price (value) and processing of the fibre (Hunter, 2020). In order to separate the fine down fibres from the coarse guard hairs, the fibres must be dehaired (Dalton and Franck, 2001). Manually separating the fine down fibres from the cashmere goat fleece (fibre coat) is a very delicate and expensive process (Wang, Singh, and Wang, 2008). Very low yielding fleeces need extra dehairing and have a higher level of coarse fibres after dehairing. The extra processing also tends to cause more fibre breakage, leading to a lower final product prices and less added value as a result of the shorter fibre lengths and hence lower yarn quality (Ansari-Renani et al., 2013).

Dehairing can be done manually or by machine (Leeder, McGregor, and Steadman, 1998). Previously, it was done manually and the dehairing process of less than 60 g fibre took up to two hours (Hunter, 2020). Later, in the 1870s a special dehairing process for cashmere, the details of which were kept secret, was developed by Joseph Dawson, the founder of Dawson International, still a major operator in the luxury fibre field (Dalton and Franck, 2001). After the first commercial dehairing machine invented by Joseph Dawson and Sons, a wide variety of patents and designs for dehairing machines emerged. Mechanical dehairing essentially consists of three phases: opening up (individualizing) of the fibres, separation and removal of the guard hair and fibre mixing. Fibre rigidity/stiffness (which is largely determined by the fibre diameter) as well as fibre length play an important role in efficient mechanical dehairing (Hunter, 2020). If the diameter ratio of coarse guard hairs and fine down fibres is less than 3.5, it becomes very difficult to separate the down and guard hair from each other based on their differences in elastic recovery and rigidity. Once these fibres become intimately mixed and intermingled, dehairing is more difficult. Therefore, production rates for dehairing are about 5 to 10% of the commercial rates for carding wool (McGregor, 2012). Commercial dehairing production rates are around 3 kg/h/m width (Hunter, 2020). During machine dehairing, the length of the fibres is reduced

(McGregor, 2018). High (>80%) relative humidity is required to control static electricity during the dehairing and subsequent processing of cashmere (McGregor, 2012). The separation criterion used for dehairing by the combing principle is the fibre length, and the relatively short fibres are removed. Therefore, it can only be used if the fibre lengths of the fine down fibres and the coarse guard hair do not overlap to any significant extent. In addition, a roller carding principle of dehairing has also been developed. It is possible to use either a modified Shirley Analyser or a Shirley Trash Separator for the laboratory dehairing of cashmere, and for determining the cashmere yield, although some caution is necessary in the interpretation and application of the results, particularly those pertaining to fibre length (Hunter, 2020).

Generally, some coarse hair still remains after machine dehairing, if the amount is 0.2% or less, it is used for knitting and if 0.5% or less it is used for weaving (Weijer, 2011). In order to reduce the coarse guard hair ratio below 0.5%, the dehairing process must be repeated many times (Leeder, McGregor, and Steadman, 1998). Coarse fibre (>30 μm) contents for Chinese Super Grade, First Grade and Second Grade cashmere are less than 0.1%, 0.2% and 0.5%, respectively (Hunter, 2020).

Increased nutrition during cashmere fibre growth increases fibre production (McGregor, 2002). Older animals produce relatively less fine down fibre and more coarse guard hair (Süpüren Mengüç and Özdil, 2014). The total hair removed from a cashmere goat can be about 800 to 900 g. Of this, some 500 g are clipped guard hair. The remaining 300 g of raw cashmere loses some 20% in mass after hand-sorting. Furthermore, the willowing process leads to the removal of sand and dirt, resulting in a 10% mass loss. The remaining approximately 200 g of fibre suffers an additional 20% mass loss during scouring. Dehairing to remove coarse hairs further reduces yield by 30%. As a result, approximately only about 100 g of cashmere (fine down) fibre remains (Hunter, 2020). In general, 100-160 grams of usable fine fibre is obtained from a goat per year (Dalton and Franck, 2001). The maximum yield is obtained from the Liaoning breed in China, which produces about 500 g/year of fibre (Vineis, Aluigi and Tonin, 2008). It is also stated that the Arbas breed of cashmere goat from Inner Mongolia produces excellent quality cashmere with high fibre yield of up to 400 grams (Hunter, 2020). Studies have shown that seasonal conditions, changes in animal weight and grazing as well as supplemental feeding have a significant effect on cashmere yield (Leeder, McGregor, and Steadman, 1998). The male goat gives a larger quantity of down fibre than the female goat (Alvigini, 1979).

In order to be able to spin yarn from fine down fibres, it is necessary to perform the dehairing process first. The yarn can then be produced in either the woollen or worsted system, depending on the fibre length and the intended end use (Dalton and Franck, 2001). Machines similar to those used for fine wool are used, although the settings, conditions, etc. are specific to cashmere and often kept secret (Hunter, 2020). When the fibres are long enough, they can be spun on a worsted, rarely semi-worsted, system (Dalton and Franck, 2001). Cashmere grown in Australia and New Zealand is suitable for processing on the worsted system since it is longer than that produced in other countries (Hunter, 2020). Only 10% of the cashmere produced in the world is processed on the worsted system (McGregor, 2002). It is stated that 120 g of fine down fibre can be reduced to 66 g at the spinning stage in Scotland (Hunter, 2020).

2.2.4. Classification of Cashmere Fibres

According to its fibre surface structure as well as handle, cashmere can be differentiated into 'cross-bred cashmere' and 'classical Asian cashmere'. Various 'cashmere fibre types', are classified according to their average fibre diameter and level of guard hair and colour. International markets generally evaluate (grade) raw cashmere quality according to fineness (fibre diameter), colour, fibre length and level of contamination (e.g., bits of skin, vegetable matter, etc.). The upper threshold of the mean fibre diameter of commercial cashmere is around 19 μm, being set at 16 (+0.5 μm) by the Chinese National Standard, at 18.5 to ± 0.5 μm by the CCMI (Cashmere and Camel Hair Manufacturers Institute) in the USA, and at 19 μm by the ASTM. First class cashmere should have a mean fibre diameter (MFD) below 15.5 μm and a soft handle. Cashmere with a MFD higher than 15.5 μm is second class. Goat fibre samples with a MFD between 19 and 23 mm are classified as "Cashgora" (Hunter, 2020).

In Mongolia, cashmere is divided into two classes: standard and superior quality. Goods consisting of 50% fine down fibres are called "standard". Goods with a fine down fibre ratio of 70-80% are defined as "superior quality" (Atav et al., 2003). According to the Mongolian standard (MNS 38-2000); a cashmere sample is classified as super quality between 13.5-15.50 μm, first quality between 15.51-16.50 μm, second quality between 16.51-17.50 μm and third quality between 17.51-19.0 μm (Phan, 2007).

There are two types of cashmere fibres in Afghanistan, the more expensive "spring (*Bahari*) cashmere" shorn from live animals in spring, and the cheaper "tannery (or skin) cashmere" obtained from the skins of slaughtered (dead) animals. The majority of the 'spring' cashmere is indeed mixed with 'skin' cashmere, often in a 30:70 or 40:60 ratio, to reduce the price of the overall product. A smaller proportion is kept separate and sold as spring cashmere for a higher price. International (mostly European) buyers prefer these two qualities not to be mixed and to pay prices dependent on quality. During 2006 the prices for bahari cashmere varied between 14 and 20 US$/kg. The price for skin cashmere is much lower than bahari cashmere. Again, depending on the quality and the yield price for skin cashmere (i.e., cashmere removed from the skin of a dead cashmere goat) varied between 9 and 13 US$/kg (Weijer, 2011).

Iranian cashmere is classified as white, grey, dark brown and black according to the colour. The raw cashmere collected from the production regions of Iran is first sent to Tehran to be classed by the experts there. An expert can make the classification of only 1-1.5 kg cashmere per day (Atav et al., 2003).

2.2.5. Microscopic Properties of Cashmere Fibres

The microscopic structures of the fine down fibres of cashmere are very similar to merino wool, with a cuticle and a cortex, and no medulla (Atav et al., 2003). The cortex of cashmere fibres is predominantly composed of ortho and meso cells, while the cortex of wool is predominantly composed of ortho-and-para-cortical cells. Mesocortical cells have an intermediate structure between ortho and para (McGregor, 2012). Roberts (1973) examined a 16.9 µm Mongolian cashmere and a 17.3 µm South African wool with a transmission electron microscope (TEM). As a result of his study, it was determined that the wool fibre consisted of 65.2% ortho and 34.8% para cortex, while the cashmere consisted of 50.4% ortho and 49.6% para cortex. In both the wool and cashmere, the average diameter of cortical cells ranged from 4.8 to 7.5 µm and their length from 84 to 111 µm. Generally, finer fibres have smaller and shorter cortical cells compared to coarse fibres (McGregor, 2012).

Coarse guard hair consists of a cuticle, cortex and medulla (Atav et al., 2003), with the medulla generally making up the major portion of the fibre. Nevertheless, the medulla is usually not seen at the root and tip of the hairs

(Von Bergen and Krauss, 1942). In coarse guard hair, the medulla can be interrupted or continuous (Dalton and Franck, 2001), with hair containing a segmented medulla is rarely being encountered (Von Bergen and Krauss, 1942). There is also intermediate (medium-thick) coarse guard hair without a medulla which is more similar to the fine down fibres than to the coarse guard hair, the tips of these fibres not being serrated. They have smoother edges, like fine fibres, but an irregular wavy mosaic pattern like the coarse fibres (Markova, 2019). The cross-sectional, longitudinal and surface (scale) views of cashmere fibres are shown in Figure 26.

The cross-sections of the fine down cashmere fibres are close to round whereas those of the coarse guard hair vary considerably both in shape and size. Although their cross-section is circular, an ellipsoidal appearance is also apparent in Mongolian cashmere. Hairs with circular cross-sections are generally of uniform diameter, with little variation in diameter over their length. In heterotype hairs, the elliptical shape occurs, more than 2/3 of these being kemp (Atav et al., 2003).

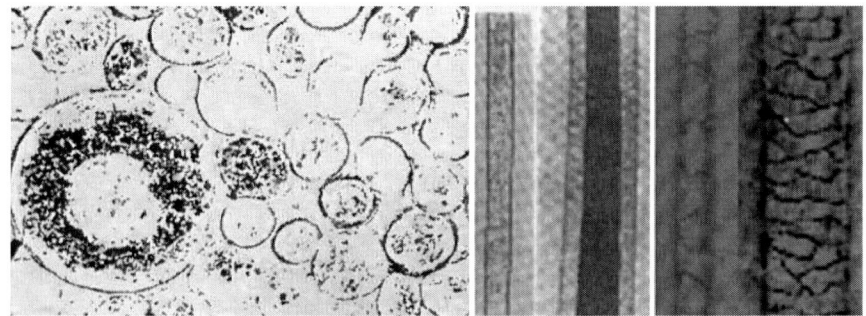

Figure 26. The cross-section (Von Bergen and Krauss, 1942), longitudinal view (fine down fibre and coarse guard hair) and the view of the scale layer (fine down fibre and coarse guard hair) (Yazıcıoğlu, 1996) of the cashmere fibres.

Cashmere fibres can have non-uniformly or uniformly distributed pigment granules in their cortex, which can be clearly seen when viewing the fibre cross-section with a light microscope. These pigment granules distinguish cashmere fibres from merino and mohair fibres that do not contain pigment granules (Markova, 2019). The distribution of colour pigments is uniform in light coloured Chinese and Mongolian cashmere, but are unevenly distributed in coloured Iranian cashmere (Atav et al., 2003). The cortical layer of white and grey hairs shows distinct longitudinal stripes, while brown hairs are completely covered with small pigments (Von Bergen and Krauss, 1942).

In cashmere fibres, scales of epidermis cells appear more prominent than in mohair fibres (Atav et al., 2003). Cashmere fibres have wider, thicker and shorter scales (approximately 85% of the scale lengths in mohair) than mohair (Hunter, 1993), but the scales are not as pronounced as in wool fibres. This feature makes cashmere more silky and smoother than wool and gives it a unique handle (Atav et al., 2003). Different types of classical Asian cashmere exhibit similar fibre surface properties and typically have a smooth cylindrical and semi-cylindrical scale shape. Each cuticle cell encloses all or half of the fibre. Cashmere fibres generally have an average scale frequency of 6-8 (i.e., the number of scales per 100 μm fibre length) and a scale height of around 0.4 μm (Dalton and Franck, 2001; Hunter, 2020), with the edges of the scales not sharp, being round and mosaic-like shape with wavy edges. The covering scales of the coarse hairs are wavy and irregular in appearance (Atav et al., 2003).

Cashmere fibres from the relatively new production areas exhibit quite different scale shapes than the Asian types, usually having a higher scale frequency (>8 per 100 μm) and visually more complex and less regular and more like mohair. They also tend to be more lustrous and smoother (slippery), with a harsher handle than their Asian counterparts (Hunter, 2020).

In a study conducted on cashmere fibres by Yang et al. (2005), the mean scale height was found to be 0.34 μm for fibres with an average fibre diameter of < 18.0 μm and 0.36 μm for fibres with an average fibre diameter of ≥ 18.0 μm. It was also found that as the average fibre diameter increased, the scale frequency decreased and the ratio of fibre diameter to scale length increased.

McGregor and Liu (2017) investigated the cuticle properties of cashmere obtained from goats exposed to controlled feeding. It was found that the scale frequency ranged from 5.8/100 μm for coarser cashmere produced by goats fed high protein diets to 8.2/100 μm for fine cashmere fibres obtained from goats raised under poor feeding conditions.

Roberts (1973) studied Mongolian cashmere using scanning electron microscopy (SEM), finding that the scale margins of cashmere fibres were not as prominent as for wool fibres. He also found that the margins of the scales were more prominent for coarse fibres than for the fine ones. He attributed the softness and lower shrinkage of fine cashmere fibres to its their less prominent surface scales compared to wool.

2.2.6. Physical Properties of Cashmere Fibres

The commercial value of cashmere fibres is determined by following parameters (Dalton and Franck, 2001):

- Fineness (diameter),
- Length,
- Clean fibre yield,
- Cleanness and
- Colour

Fineness (Diameter)

Although the term fibre diameter suggests that the fibre has a circular cross-section, in reality the cross-section of most animal fibres is elliptical rather than circular. The term ellipticity is used to describe the deviation from the circular shape and is commonly calculated as the ratio of the major axis to the minor axis of the fibre ellipse (the ratio between large diameter and small diameter) (Hillbrick, 2012). Hillbrick (2012) found that the ellipticity of cashmere fibres is lower than that of wool fibres with a similar fibre diameter, that is, cashmere fibres are significantly more circular. The ellipticities of cashmere and wool fibres were found to be 1.14 and 1.19, respectively. In contrast to this, the ellipticity of cashmere fibre was found to be 1.2 by McGregor and Lui (2017). According to McGregor and Quispe Peña (2017), the ellipticity of Chinese, Iranian and Australian cashmere was as 1.18, 1.21 and 1.19, respectively.

The diameters of the fine down cashmere fibres vary from about 13 to 19 μm (Vineis, Aluigi and Tonin, 2008), with that of the coarse hairs varying between 30 and 150 μm, with an average of around 60 μm. The diameter of the fine down fibres from cashmere goats is similar to that from common goats (about 14 μm), although cashmere goats normally produce finer down fibres than common (ordinary) goats (Hunter, 2020). Single fibre diameters range from 8 μm to 24-25 μm in cashmere fibres subjected to a good dehairing process (Dalton and Franck, 2001) and fibres coarser than 22 or 23 μm are rarely encountered. Fibres with a fineness between 25 and 30 μm are morphologically conspicuous and in appearance more like the coarse guard hair (Hunter, 2020).

Cashmere fibre diameter is determined by both genetic and environmental factors ("Heritage Cashmere", 2003), fineness and quality varying according to the region where the fibre is produced. The best cashmere is obtained from goats living north of the 40th parallel in Inner Mongolia (Göktepe, Canipek, and Soysal, 2018). Cashmere goats bred in milder climates do not produce fibres as fine as they do in their natural colder habitat, but the fibres are still very soft compared to many other animal fibres (Dalton and Franck, 2001). Cashmere is at its finest in the first year of a goat's life. The mean fibre diameter of Mongolian cashmere tends to increase with goat age, by, on average, 1 μm per year from 1 to 5 years of age (Hunter, 2020). The best cashmere is obtained from the neck of the goat (Karthik, Rathinamoorthy, and Ganesan, 2015).

Chinese cashmere, with a mean fibre diameter of 14-16 μm, is considered the best quality, Mongolian cashmere being slightly coarser than that having a mean fibre diameter of 16-17.5 μm. Due to crossbreeding for higher yields, some Mongolian cashmere increased in diameter, with an associated quality loss (Dalton and Franck, 2001). The mean fibre diameter of Mongolian cashmere increased from about 16 μm to between 17 and 19 μm after the mandatory culling of older male goats was revoked. However, the fibre diameter was reduced to about 15.5 μm with the introduction of the Liaoning breed in Inner Mongolia (Hunter, 2020). Iranian and Afghan cashmere range from 16-19,5 μm; New Zealand and Australian cashmere from 16-18.5 μm (Dalton and Franck, 2001). Indian cashmere is finer than 16 microns ("Diamond Fibre Pashmina", 2009). The diameter of the fine down fibres of Kyrgyz cashmere varies between 13 and 16 μm, with the diameter of the coarse guard hairs varying between 70-90 μm (Kerven and Toigonbaev, 2010).

The quality of cashmere, especially its fineness and length, plays an important role in determining its end use (product). For knitted garments, where softness is of primary importance, higher quality fine cashmere is preferred, while coarser (lower quality) fibres are usually used in weaving (Hunter, 2020). Fineness has a significant effect on price (Dalton and Franck, 2001), the finer the fibre, the higher the perceived quality and price (Hunter, 2020). For example, Iranian and Afghan cashmere is 2-3 μm coarser than Chinese cashmere and some 40-50% cheaper (Dalton and Franck, 2001).

Length
The length of the cashmere fine down fibres varies between about 2 and 5 cm (20 mm and 50 mm) on average (Vineis, Aluigi, and Tonin, 2008), while the lengths of the coarse guard hair vary between about 5-12.5 cm (50-125 mm) (Cook, 2001). Cashmere fibres that have undergone a good dehairing process

typically have a CV of diameter of 20% (even up to 25%). The mean length of the fine undercoat fibres of raw Chinese cashmere typically varies from about 21 to 40 mm, the lengths of the individual fibres ranging from about 5 to 80/90 mm, depending on the quality of the sample. The average fibre length of dehaired Chinese cashmere ranges from about 24 mm for the poorer qualities (e.g., Brown, Second Grade) to 36 mm for the top qualities (e.g., Super Grade), the average fibre length of cashmere usually being over 38 mm (50% of the fibres being over 40 mm) for two provinces in Mongolia that gave the best results. The mean length of Asian cashmere is between 21 and 40mm, with individual fibres ranging in length from 10- and 90-mm. Super Class cashmere has a fibre length of 38 mm and First Class 36 mm. Cashmere grown in Australia and New Zealand tends to be longer than that grown in other countries (Hunter, 2020). The coarse guard hairs are shorter than the fine down fibres in Australian cashmere goats (McGregor, 2012). In the study by McGregor (2001) the fine down fibre length was found to be 75.3 ± 20.4 mm and the coarse guard hair 63.3 ± 27.0 mm for Australian cashmere. The length of the fine down fibres of Kyrgyz cashmere varies between about 40 and 50 mm and that of the coarse guard hairs varies between about 150 and 170mm (Kerven and Toigonbaev, 2010).

Fibre length is important in determining the application and use of cashmere (weaving or knitting) and affects the quality of the garment, e.g., the low fibre length increases pilling in knitted clothes. In addition, the dimensional stability of products made from Mongolian cashmere, which is coarser but longer than Chinese cashmere, is higher (Weijer, 2011). It is desirable for very high-quality knitting and weaving yarns that the cashmere mean fibre length should be 40 mm or even longer (Hunter, 2020).

Strength

In one study, single fibre strength values (tenacity) were found to vary from 7.0-15.5 cN/tex for Australian cashmere (MFD range 15.1-17.9 μm), 8.2 - 11.2 cN/tex for two samples of Chinese cashmere (MFD 13.6 and 13.9 μm), and 16.2 cN/tex for an Iranian cashmere (MFD 17.4 μm) (McGregor, 2018). For Mongolian cashmere, it is stated that the single fibre strength is 13.69 cN/tex (Hunter, 1993). Roberts (1973) found that Mongolian cashmere and lamb's wool with similar average fibre diameters have similar stress-strain properties, elongation at break, and tensile strength. Hillbrick (2012) found that cashmere fibres have a higher Young's modulus than wool of equivalent diameter. McGregor and Postle (2004) found a bundle fibre strength of 9.9 cN/tex for cashmere fibres from various origins. In a study on Mongolian cashmere, the bundle fibre strength was found to be approximately 12.5 cN/tex (Badmaanvambuu et al., 2003). Gupta et al. (1992) found the strength of cashmere guard hair to be 11.6 cN/tex.

Elongation (Elasticity)
McGregor and Postle (2004) found the fibre bundle elongation at break to be 40.8% for cashmere fibres from various origins. In a study conducted on Mongolian cashmere, the bundle fibre elongation at break was found to be 45% (Badmaanvambuu et al., 2003). Gupta, Arora and Patni (1992) found the elongation at break of cashmere guard hair to be 34%.

Bending Stifness
Tester (1987) found that cashmere fibres have, on average, fewer cuticle cells around the fibre cross-section than 16-18 μm Merino wool fibres. It has been stated that the lower cuticle cell thickness in cashmere fibres may be related to the lower bending stiffness of cashmere fibres compared to Merino wool (McGregor, 2018).

Clean Cashmere Yield
The clean cashmere yield of cashmere goat fleeces depends on the scouring yield which depends on the cleanliness (foreign substance content) of the fleece and the down fibre yield (down fibre/coarse hair ratio) in the subsequent dehairing process. According to this, clean cashmere yield (%) can be expressed as follows (McGregor, 2012):

Clean cashmere yield of greasy fibre (%) = washing yield (%) X cashmere yield in clean raw fibre (%)

(1)

Table 13. Average scouring yield of cashmere fibres of different origins (Hunter, 2020)

Cashmere Origin	Average fibre diameter (μm)	Average scouring yield (%)
Chinese	14-16	75
Mongolia	16-17	75
Iranian	17-19	80
Australia	15.5-18.5	92

Raw cashmere contains between 25% and 30% sand and dust, 4-5% grease and 65-70% clean fibre (Hunter, 2020). The grease content of Australian raw cashmere fibres averages 2.5% and is very low compared to Chinese and

Mongolian raw cashmere fibres which contain an average of 5% grease (Wang, Singh and Wang, 2008). The proportion of dust and sand in the foreign materials found in cashmere is high, and that of vegetable foreign materials low. The amount of foreign matter that can be extracted with solvents is generally more than 3% (Atav et al., 2003). McGregor (2004) found the vegetable matter content of Australian cashmere to be 0.9%. Scouring yield in cashmere is about 70%, the average for cashmere from different origins is given in Table 13 (Hunter, 2020). As can be seen from the table, the scouring yield for Australian cashmere is higher than that of the others. McGregor (2002) stated that Australian cashmere is very clean compared to Chinese cashmere, with an average scouring yield of over 96% and a soil content of less than 2%.

Yield is calculated for cashmere in terms of the ratio of the weight of fine down fibre obtained after dehairing to the total weight of the greasy fibres (Leeder, McGregor and Steadman, 1998). The fine down fibre yield of dehaired cashmere ranges from about 30% to 70% of the raw fibre weight. This largely depends on the fibre harvesting method (i.e., combing or shearing) and the type and quality of the goat. For example, the dehairing of shorn fibres is more difficult than of combed fibres (Hunter, 2020). In traditional cashmere producing countries, such as China and Mongolia, where low-cost workforces are available, fine down cashmere fibres are hand-combed from each goat, and the coarse protective outer hair content is relatively low. For example, prior to dehairing, fine quality Chinese cashmere contains only 18% coarse guard hair. In Australia, since the fibre is obtained by shearing, the coarse guard hair content is much higher (around 70%) (Wang, Singh and Wang, 2008). In a white coloured super quality Mongolian Cashmere sample, there is generally 88% fine down fibre and 12% coarse guard hair (Atav et al., 2003). In Table 14, down fibre/guard hair ratios and moisture, grease and water-soluble substance contents found in a study for Chinese and Australian cashmere are given (Hunter, 1993).

Table 14. Moisture, oil and water-soluble matter content of Chinese and Australian cashmere (Hunter, 1993)

Fibre	Area	Yield (%)	Fibre Diameter (μm)	Moisture (%)	Surface Grease (%)	Internal Lipids (%)	Water Solubles (%)
Chinese cashmere	Down fibre	87	15.3±3.0	12.7	2.0	2.1	1.5
	Guard hair	13	49.3±8.3	12.1	0.8	-	-
Australian cashmere	Down fibre	19	15.6±3.5	13.6	2.6	1.1	0.9
	Guard hair	81	72.5±28.0	13.5	1.0	4.8	0.3

As can be seen from the Table 14, down fibre yield in Chinese cashmere is high compared to Australian cashmere. McGregor (2002) stated that the down fibre yield of Australian cashmere is between 30 and 35%.

Crimp

Cashmere fibres usually have only a little crimp (i.e., a low crimp frequency). Cashmere from the newer growing regions, such as Australia, New Zealand and the USA, has an even lower degree of crimp than cashmere from the traditional sources, such as China, Mongolia, Iran and Afghanistan. It has been found that fibre crimp in Australian cashmere, as in wool, has a negative relationship with fibre diameter and is influenced by the diet of goats (Hunter, 2020). It is also reported that cashmere fibres containing more mesocortical cells and have a higher microfibril packing density and arrangement compared to wool fibres of the same average fibre diameter, which may be associated with the lower crimp of cashmere (McGregor, 2018). The crimp frequency of cashmere varies between 1.5 and 7 crimps/cm, and the curvature (degree of crimp) varies between 27 and 84 °/mm (Mcgregor, 2007). McGregor (2001) found the crimp frequency for Australian cashmere to be 3.2 ± 0.9 crimps/cm. For Chinese cashmere, the goats age and sex affect fibre crimp, curvature being 52°/mm for male goats, 65°/mm for female goats and 78°/mm for kids. For Mongolian goats, on the other hand, it was found that the curvature ranged from 55 to 66°/mm and decreased with an increase in fibre diameter. In addition, it has been found that there are differences in the degree of crimp between certain regions (Hunter, 2020). Various crimp shapes have been reported for wool and cashmere, ranging from helical to sinusoidal or planar. Crimp density and crimp shape influence fibre cohesion, which is important for processing staple fibres. Although processing, particularly mechanical, reduces fibre crimp, about half of the crimp is recovered during steam relaxation (Hillbrick, 2012).

Colour

In China and Mongolia, as well as in Iran and Afghanistan, dehaired cashmere fibres can have various colours, including white, light grey, dark grey and shades of brown, while cream, light brown and dark brown are typical colours of Iranian and Afghan cashmere (Dalton and Franck, 2001) (Figure 27). There is also black cashmere (Hunter, 2020). These and other colours probably come from cross breeding in the past with other types of goats (Alvigini, 1979). There are differences between the lightness and yellowness of white cashmere from different countries, and this is also influenced by the breeder (McGregor,

2012). As the fleece weight of goats increases, the lightness of the colours of the fibres produced increases and the yellowness decreases. The reason for this is explained by several different mechanisms, such as a decrease in the average UV irradiation to which cashmere fibres are exposed as the fleece production (weight) increases, the relative dilution of the natural chromophores in the fleece as the fleece growth increases, or a change in the fibre reflectance properties related to the scale size and surface properties (McGregor, 2012).

Colour has a significant effect on price, white being the most valuable because it can be used not only as it is but can be dyed to the pastel shades which are often required for knitwear. Brown cashmere is the least valuable because it can only be dyed to dark shades. Chinese cashmere is predominantly white (Dalton and Franck, 2001), whereas only very small amounts of Afghanistan cashmere are white, most being predominantly dark in colour. Mongolian cashmere is lighter in colour than Afghanistan cashmere. Therefore, the average price of Afghanistan cashmere is lower than that of Chinese and Mongolian cashmere. For example, in 2006, the average price of Mongolian raw cashmere at the farmer level was $23/kg, while in Afghanistan it was around $16/kg (Weijer, 2011). The presence of even as few as 5 coloured fibres in 1g of white cashmere affects the quality negatively (Hunter, 2020).

Figure 27. White and coloured cashmere fibres (adapted from Atav et al., 2003).

The analysis of recent commercial cashmere prices indicates that within certain colour ranges processors pay premium prices for finer cashmere, regardless of colour which means that main differences in the price of raw cashmere are based on fibre diameter rather than colour. Though white cashmere is more valued by processors as it can more readily be dyed, there is only a one-dollar (3%) difference in the price of white and brown cashmere of superior quality (13.0-15.5 micron) of Mongolian cashmere (see Table 15).

However, there is 12 dollars difference in the price of white superior grade cashmere and the coarsest white cashmere (over 17.6 micron), a difference of over 50% (Ansari-Renani et al., 2013).

Similar to the example given above, some indicative prices paid for dehaired cashmere by major UK processing company are shown in Table 16. As can be seen from these prices, international processors pay according to quality, which is primarily determined by the fineness of the fibre, i.e., fibre diameter in micron (Ansari-Renani et al., 2013).

Table 15. Price ($/kg) difference per quality in Mongolia (Weijer, 2011)

Colour/Diameter	Superior (13.0-15.5 micron)	Grade I (15.51-16.8 micron)	Grade II (16.81-17.6 micron)	Grade III (17.61-19.0 micron)
White	34	31	29	22
Light grey	34	31	29	22
Grey	33	31	28	21
Brown	33	30	28	21

Table 16. Prices paid for dehaired cashmere by major UK processing company 1992-2002 (Ansari-Renani et al., 2013)

Diameter (Micron)	Colour	Price (USD/kg
15	White	50-140
15	Coloured	45-130
16	White	50-130
16	Coloured	45-120
17	White	40-115
17	Coloured	35-100

Other Properties

Horikita et al. (1989) found that the densities of white and brown cashmere fibres were 1.308 and 1.311, respectively. Cashmere fibres are generally not as lustrous as mohair (Hunter, 2020).

Cashmere does not cause allergies, does not irritate the skin, does not prickle, and is a silk-like but warm fibre. Sweaters made of cashmere provide excellent comfort. Beyond its softness and elasticity, cashmere is also stated to provide better heat insulation (warmth) than wool ("Women's Cashmere", 2011). Cashmere, as in the case with other animal fibres, provides a buffer to sudden changes in temperature and humidity thereby adding to its comfort, which makes cashmere an ideal fibre, especially for baby clothes ("Silk and Cashmere", 2002).

In the literature, it is stated that there is no significant difference in the friction properties of the scale surface of cashmere and wool fibres when measured at the nanoscale, but at the macro scale, the fibre-metal friction coefficient and the directional friction effect (DFE) of the cashmere fibre (i.e., against scale and with-scale friction) is significantly lower than that of wool of a similar diameter (Hillbrick, 2012). Liu and Wang (2007) examined the felting tendency of alpaca, cashmere and wool fibres. They found that short and fine cashmere fibres had a lower felting tendency than wool fibres with a similar fibre diameter. In addition, they found that the fibre length had a significant effect on felting, with the longer fibres felting more than the shorter fibres.

McGregor and Schlink (2014) investigated the felting properties of cashmere from different origins. Research on the felting properties of animal fibres, such as cashmere, quivit, camel hair, llama, guanaco, bison wool, bovine fibre and yak wool from different production areas, showed that there were great differences in the felting properties of these fibre types and even within a fibre type as well, with the felt ball densities of cashmere and cashgora fibres from certain origins being greater than those of the other fibres. It was found that the densities of the felt balls decreased in the order of Chinese cashmere < Iranian cashmere < Australian cashmere. Cashmere goats were also given different diets and it was found that the cashmere produced by the poorly fed goats had a lower felting tendency than the cashmere produced by the better fed goats. The felting tendency also increased when cashmere was blended with superfine wool with a high fibre crimp. However, when the same cashmere was blended with a low-crimp superfine wool, the felting tendency changed very little or not at all. McGregor and Liu (2017) also found that cashmere produced by malnourished goats showed a lower tendency to felt than cashmere produced by better fed goats. This was related to differences in fibre crimp, although the scale frequency was higher.

The resistance to compression of cashmere fibres is affected by the origin of the fibres (McGregor, 2012). Cashmere from new origins, mainly Australia, had a low resistance to compression. This indicates that these cashmere fibres are more compressible, i.e., softer, than those from traditional cashmere producing countries (McGregor, 2004). The softness of cashmere has been compared to the softness of wool and it was found that coarse (19.8 μm) cashmere was softer than 17.2 μm wool (Hillbrick, 2012).

McGregor (2007) suggested that the softness of cashmere is associated with its monoplane crimp form and low resistance to compression resulting from the low fibre crimp frequency. He noted that the shear modulus is significantly different for the various keratin fibres, cashmere fibres having a much lower shear modulus than wool of similar diameter. The main factor affecting the shear modulus is the fibre matrix, and the structural differences in the matrices of the various animal fibres could at least partly explain the observed differences in their softness.

2.2.7. Chemical Properties of Cashmere Fibres

The chemical properties of cashmere fibres are very similar to that of wool fibres (Cook, 2001). Compared to wool fibre, which on average contains 3.7% sulphur and 16.5% nitrogen, cashmere has slightly lower sulphur and nitrogen contents of 3.4% and 16.4%, respectively (Hassan, 2015).

Roberts (1973) compared the amino acid composition of Mongolian cashmere with a diameter of 16.9 microns and South African wool with a diameter of 17.3 microns and concluded that the amino acid composition of the two fibres was very similar. Only cystine, tyrosine (12% more in cashmere than in wool) and pyroline (9% less in cashmere than in wool) of the amino acids differed significantly. He also observed that polar amino acids such as serine and threonine, were more abundant in cashmere than in wool. The amino acid composition of cashmere is influenced by the diet of the goats, the type of feed (high quality high protein feed or pasture grazing) and country of origin. In addition, it is stated that there are significant differences between the amino acid composition of the fine down fibres and coarse guard hair of the same cashmere goat (McGregor, 2012).

McGregor and Tucker (2010) found that there were differences in the amino acid composition of cashmere down fibre and guard hair, but no significant difference in cystine content. Variations in the amino acid composition of cashmere fibres affect both their physical and chemical reactivity and properties (McGregor, 2012). The amino acid composition (μmol/g) of cashmere fibres of different origins are given in Table 17 (McGregor and Tucker, 2010).

As can be seen from Table 17, the tyrosine and phenylalanine contents of Australian cashmere are lower than those of Chinese cashmere. Feeding with protected protein increases the tyrosine and phenylalanine content of cashmere. If the same mechanisms, that operate in wool, also operate in cashmere, then Chinese cashmere and cashmere grown from goats fed high levels of protected protein may have a greater propensity to yellow when exposed to UV light. This may explain the finding that the origin of cashmere explains over 50% of the variation in the colour of white cashmere, and that processed Chinese white cashmere had lower lightness and greater yellowness compared with Australian white cashmere (McGregor, 2018).

Table 17. Amino acid composition (μmol/g) of cashmere fibres (McGregor and Tucker, 2010)

Amino acid	Chinese Cashmere	Iranian Cashmere	Australian Cashmere
Alanine	482	472	472
Arginine	575	554	562
Aspartic acid	475	468	468
Cystine+Cysteine	425	394	411
Glycine	696	644	571
Glutamic acid	1207	1183	1189
Histidine	72	68	70
Isoloysin	256	259	261
Lysine	219	214	210
Leucine	607	583	578
Methionine	34	30	31
Phenylalanine	233	223	219
Pyroline	692	688	715
Serine	903	843	836
Tyrosine	325	299	255
Threonine	504	503	527
Valine	443	432	433

Cashmere fibres are less resistant to chemicals than wool (Cook, 2001). Washed cashmere normally contains 12% moisture (Atav et al., 2003), and wet out faster with water (i.e., have a greater wettability) than wool (Cook, 2001). Cashmere fibres wetted with water become saturated in a few seconds, while wool fibres take longer to reach this state (Atav et al., 2003). Roberts (1973) stated that the surface of cashmere fibres is more hydrophilic than that of wool and that the easier wettability of cashmere fibres is due to this hydrophilic structure. He added a certain amount of paraffin oil to wool and cashmere fibres and washed these fibres in soap solution under the same conditions. As a result of this process, more oil remained on the wool fibre. Based on this, he explained that the surface of cashmere fibre is more hydrophilic than that of wool.

In a study by Liu, Xie, and Liu (2017), it was found that when treated with either hydrochloric acid or sodium carbonate, the strength of cashmere fibres decreased, with the percentage decrease in fibre strength increasing with increasing acid or alkali concentration (between 1 and 4 mol/L). In addition, it was found that the fibre strength decreased more after the treatment with sodium carbonate than after the acid treatment at the same concentrations. Since disulphide or salt bridges in the structure of keratin can be broken down by alkali, cashmere is resistant to acid, but not to alkali. Cashmere fibres are more sensitive to alkali than wool (Hunter, 2020), being easily damaged by alkalis, such as soda ash used in wool scouring, and it dissolves completely in caustic solution (Cook, 2001).

Because cashmere fibres have fewer cuticle cells than wool, it seems likely that photodegradation is a more serious problem for cashmere than for wool. In one study, it was found that goats grazing outdoors on pastures produced less lustrous and more yellow cashmere than those fed indoors (under cover). It is thought that the yellowish colour is related to the breakdown of amino acids in the structure of the fibres as a result of damage by ultraviolet (UV) rays (McGregor, 2012). In addition, certain researchers have detected cysteic acid in cashmere samples, although the amount is quite low (9-17 mmol/g). This indicates that some photochemical degradation occurs during the growth of the fibres (McGregor, 2018).

It was originally thought that the down fibres (cashmere) were protected by the guard hairs from sunlight and other harmful environmental factors and therefore not subject to photochemical degradation. Nevertheless, studies have revealed that cashmere fibres are also weathered and that cashmere from different locations are exposed to atmospheric conditions to different extents. Weathering refers to the degradation of animal fibres that occurs during their growth, due to exposure to sunlight, water, and air. Exposure of cashmere to weathering decreases its tenacity and elongation, increases its yellowness, and reduces the brightness of white cashmere (McGregor, 2018).

2.2.8. End-Uses of Cashmere Fibres

Cashmere is one of the most sought-after textile raw materials since ancient times. It is soft and suitable for making elegant, stylish, attractive and most comfortable clothing (Atav et al., 2003). The largest portion of cashmere fibre produced in the world is used in the production of fine woollen yarn for fashionable knitwear in Scotland. In woven fabrics, both woollen and worsted yarns are used (Dalton and Franck, 2001).

Cashmere fibres can also be blended with other fibres, and these blends generally contain at least 10% cashmere (Atav et al., 2003). Because cashmere is very expensive, and for better processing performance, and yarn and fabric quality, it is often blended with fine wool, such as lamb's wool, or silk, and sometimes also with a small proportion of nylon, to provide additional stability and strength. Nevertheless, when blended, its handle and softness may suffer unless it is blended with an extremely fine fibre (Hunter, 2020). Cashmere is often blended with fine merino wool to produce fabrics with a soft handle that are less expensive than 100% cashmere but still have the image of cashmere (Dalton and Franck, 2001). In terms of spinning, it is stated that a 50:50 mixture of cashmere and wool is the most suitable (Raja et al., 2011).

Cashmere products (shawls, scarves, sweaters and cashmere fabric for men's suits) have two distinctly different markets; the high-end market, dominated by European companies holding brand names, and the medium- and low-end market, dominated by China. Italy and the UK play a leading role in the branding, marketing and retail of the high-end garments, while the US is a major importer for cashmere, particularly within the medium- and lower-end markets (Weijer, 2011).

With such a high-price and excellent-image it is hardly surprising that certain unscrupulous dealers will fraudulently misrepresent the presence and/or amount of cashmere in a product. It has been estimated that 15% of garments claiming to contain real cashmere and blends of cashmere are fraudulently labeled. The policy of the Cashmere and Camel Hair Manufacturers Institute (CCMI) in this regard is to purchase products in the open market and analyse them to determine whether the fibre content is in accordance with the label on the garment (Dalton and Franck, 2001). Since cashmere is very expensive, very popular and available in rather limited supplies it is sometimes recycled or reprocessed and reused, in such cases it should, however, always be labeled accordingly as "recycled cashmere" (Hunter, 2020).

Of all clothing or home textiles bearing the "Mongolian Cashmere - made with" and "Mongolian Cashmere - pure" labels created by the Mongolian FibreMark Society, at least 50% or more, contain 100% (pure) Mongolian cashmere (Figure 28). Products bearing this mark are reliably labeled according to the origin and purity of the cashmere from which they are made and are not made from recycled or waste fibre ("Mongolian Fibremark", 2020).

Figure 28. "Mongolian Cashmere - made with" and "Mongolian Cashmere - pure" labels ("Mongolian Fibremark", 2020).

Cashmere shawls, scarves (Figure 29) and covers are some of the textile products that have enjoyed a long worldwide reputation. In the past, items such as shawls, scarves and belts were made from cashmere. Luxurious cashmere woven fabrics were used to make men's loincloths in Egypt, women's jackets in Iran, men's and women's coats (Figure 29) in Turkestan, and sashes in Tibet (Maskiell, 2002). Today, cashmere is mostly used in women's dresses and trousers. Among the most expensive fabrics included in this category, are cashmere/silk velvets, with a silk warp and a cashmere weft. Although not as common is cashmere also used in the production of men's jackets and overcoats (Atav et al., 2003). Cashmere fabrics have a nice drape (Cook, 2001), but tend to have a lower dimensional stability than wool fabrics (Corbman, 1983).

Figure 29. Scarf ("Burberry Cashmere", 2011) and women's coat made of cashmere fibres ("Cashmere Women's Coat", 2011).

Until relatively recently, the many manufacturers of high-priced cashmere garments were in Scotland and Italy, but in the late 1980s China and Mongolia set up their own spinning and knitting mills using their own dehairing technologies developed in collaboration with Japan. The knitwear they produced was less expensive, although not of such a good quality as the Scottish and Italian products (Dalton and Franck, 2001).

The cashmere market has been disrupted since luxury markets, such as it, depend not only on good marketing and product quality, but also on some degree of rarity. In addition, as the supply of cashmere has become limited, Scottish and Italian producers have struggled to procure sufficient amounts of quality raw material (fibres). In the face of this, European manufacturers have established joint ventures with partners in China. This has largely solved the raw material supply problem, and the quality has improved due to associated improvements in Chinese production methods. Scottish manufacturers, on the other hand, protect their own markets by increasing their product range and developing their retail networks. Apart from these countries, it is known that there are companies operating in the USA in the field of cashmere fabric and clothing production (Dalton and Franck, 2001).

Traditionally, cashmere garments and fabrics have been mainly produced in the UK and Italy, but there are also companies in the US that specialise in this niche market. Although cashmere garments are sold all over the world, the main markets, as with many other luxury products, are America, Japan and Western Europe. While less important than knitwear, woven cashmere fabrics for men's and women's outerwear, also represent an important market, the main markets once again being America, Japan and Western Europe (Dalton and Franck, 2001). The size of the global cashmere clothing market was $2.66 billion in 2018, the distribution by product being as follows; sweater and coats the highest share (approx. 50%), followed by trousers at around 25% and tee and polo neck shirts at around 15% ("Grand View Research", 2019).

The coarse guard hairs (outercoat) of cashmere goats are often too coarse for textile use, although raw goat hair (i.e., a mixture of fine down fibres and guard hairs) has been used in the manufacture of certain fabrics (e.g., tents) and carpets in the past (Hunter, 2020). The coarse guard hair of cashmere goats is also used in the production of ropes, sacks, rugs, carpets and under-carpet felts (Dalton and Franck, 2001; Başer, 2002).

Although machine washable cashmere garments have been produced, cashmere garments are usually cleaned by dry cleaning or gentle hand washing, using not very hot water and a pure powder or liquid soap or detergent specifically recommended for cashmere. It is recommended to use soft water for foam formation, which is important in the washing process of cashmere knitted clothes. During washing, the garment should not be rubbed or wrung and should be rinsed thoroughly in cold water, without the use of fabric softener, and then dried flat, protected from direct sunlight (Hunter, 2020).

2.3. Fibres of Goat Hybrids

Diverse and other goat-fibre alternatives, such as Cashgora, Pygora, Nigore and Pycazz, originate from crossings (and sometimes second and more crossings) that involve mohair, cashmere, or both types of goats. There are a number of goat hybrids as a result of crossing various common goats with angora goats or cashmere goats. For example, Pygora goats originated through crossing from Angora goats with the Pygmy goats. The crossings between pygora and cashmere goats are called Pycazz. On the other hand, Nigora goats come about through the crossing of Angora goats with Nigerian Dwarf goats (Ekarius and Robson, 2011). Here information is given for Cashgora, a goat hybrid having both parents as a source of luxury fibre.

2.3.1. Cashgora Fibres

Cashgora has been labeled as the first new natural textile fibre of the last 100 years (Shamey and Sawatwarakul, 2014). However, the market's initial interest in this new fibre has not continued.

A significant amount of scientific research has been carried out on the possibilities and advantages of crossbreeding between different goat breeds, with Cashgora being produced by crossing Angora goat with another goat species, such as feral or wild goats. There however, is no clarity or certainty about what kind of goat it could be wild goats, cashmere goats, Anglo-Nubian and dairy goats in New Zealand. It is therefore not surprising that the fibre has a mixed reputation in the industry, as its characteristics are unlikely to be constant with such a potential variety of parents (Dalton and Franck, 2001). Cashgora is normally obtained in the first and second such crosses, and the finest fibre is harvested in the first crossing. It is stated that Cashgora fibres can be described as "fine mohair" or "coarse cashmere" (Shamey and Sawatwarakul, 2014). Casgora fibre, which constitutes the third group of goat fibres after mohair and cashmere, is obtained from crosses of pure Angora goats with cashmere goat, dairy goat, common goat or meat goat breeds. It was first observed in Australia in 1981 that the fibres of goats obtained by crossing Australian wild goats (feral goats) and Angora goats differ from cashmere and mohair fibres, and these new fibres were named Cashgora, a word formed by combining the words cashmere and Angora (Dellal et al., 2010). It is stated that Cashgora goats (Figure 30) are better adapted to extreme winter conditions and produce more fibre than domestic goats (McGregor, 2012). Cashgora is the name given to fibres similar to superfine mohair fibre obtained by crossing Angora goats with Australian or New Zealand wild goats (Vineis, Aluigi and Tonin, 2008). Since Cashgora is coarser than cashmere, its commercial value is correspondingly lower (Bolat, 2006).

Figure 30. Cashgora goat (adapted from "Cashgora Goat", 2010).

Cashgora attracted considerable industry attention in the last half of the 1980s, and probably for this reason, in 1988, "Cashgora" was adopted by the International Wool Textile Organization as a generic term for fibres obtained from a mohair-cashmere hybrid (Dalton and Franck, 2001). Cashgora, as an animal fibre or "wool", falls within the scope of the International Wool Textile Organization (IWTO) rules regarding the commercial transactions of "wool". Cashgora also has been added to the UK's official list of 'textiles ingredient regulations' to allow Cashgora to be included on ingredient labels. In June 1988, IWTO included Cashgora in the IWTO Blue Book for the first time (McGregor, 2012).

Although the word "Cashgora" is a name used for fibres obtained from a mohair-cashmere hybrid, calling a fibre "Cashgora" does not always mean that it is a fibre obtained from the said "hybrid goats". Sometimes, fibres that resemble cashmere fibres but do not provide cashmere quality (they are generally 19-23 microns and coarser than cashmere) are also called "Cashgora" (Ekarius and Robson, 2011).

2.3.1.1. Historical Background of Cashgora Fibres

The name "Cashgora" was first used by Ms. J. Maddocks in 1972 to describe goats with an average diameter of down fibres greater than 18 µm (McGregor,

2012). However, crosses of the Cashgora-like Angora goat with two-coated cashmere goats (as well as Anglo-Nubian and dairy goats) and crosses of domestic Kyrgyz goats from the former Soviet Union and Angora goats from Turkey were already being bred in 1820 (Hunter, 2020).

2.3.1.2. World Production of Cashgora Fibres

Cashgora fibres were produced commercially in small quantities around the world, especially in New Zealand and Australia (Hunter, 2020). In countries, such as England and Germany, systems for Cashgora fibre production have been studied for many years, albeit at a limited level (Dellal et al., 2010). Cashgora production was approximately 50 tons in 1986 (Phan and Wortmann, 2000). In 1988, New Zealand produced 200 tons of Cashgora. Production was 140 tons in 1990/91, accounting for more than half of goat fibre exported from New Zealand (Hunter, 1993). Cashgora is now grown only in Kazakhstan, the Russian Federation, and other Central Asian countries where the former Soviet Union once influenced livestock breeding practices (McGregor, 2012).

While world production was estimated at 200 tons in 1990, it decreased to 60 tons in 2000 (Dalton and Franck, 2001). According to 2007 data, its annual production was 200 tons, and while its production is stable in Central Asia, it decreased in Russia (McGregor, 2012). In April 2001, the price for 20 µm Cashgora was $45/kg (Dalton and Franck, 2001). In 2007, the price of clean fibre was $8-20/kg (McGregor, 2012).

2.3.1.3. Harvesting Cashgora Fibres and Factors Affecting Cashgora Yield

Cashgora is normally obtained from the first and second crossing and is considered as "fine mohair". In the kid and young goat stage (up to 2 years of age), a Cashgora fleece contains cashmere-like fibres as well as mohair type fibres. As the animal ages, the fine cashmere type fibres are lost and the fleece characteristically turns into a type of superfine mohair, some guard hair always being present (Hunter, 1993). Cashgora has a double layer (coated) fibre structure (Shamey and Sawatwarakul, 2014). In goats producing Cashgora fibre, primary follicles produce the coarse outer hairs and secondary follicles the fine down fibres (Dellal et al., 2010). Depending on where in the world, and under which system, it's being classed, Cashgora either contains three types of fibres (down, guard hair, and intermediate hair) or consists of what might alternatively be called strong cashmere, with down (the Cashgora portion) and coarse guard hair but no intermediate hair (Ekarius and Robson,

2011). Fibre production per animal is ≈2-4 kg/year (Hunter, 2020). Usable annual fibe yield per goat ranges from under 4 ounces (113 g) to 14 ounces (397 g) (Ekarius and Robson, 2011).

Cashgora is shorn twice a year, after which the fleece must be separated, that is, the fine down fibres must be mechanically separated from the coarse guard hairs. Cashgora is classified using the criteria valid for cashmere, and fibres coarser than 30 μm are classified as guard hair. The fine down fibres constitute approximately 50% of the fleece mass /weight (Hunter, 1993). The Australian Cashmere Marketing Corporation defined Cashgora as "fleece having coarse guard hair, fine crimped down and longer shiny straight "intermediate" fibres - i.e., three fibre components". Since the fibre diameter distributions of such intermediate fibres, essentially fine guard hairs, overlap with that of the down fibres, they cannot be easily dehaired from the fine downy undercoat, which causes commercial losses. This type of fleece is lower in quality than cashmere, as it is difficult to dehair, and is therefore of lower commercial value (McGregor, 2012).

After being sorted, Cashgora fibres are usually subjected to a similar processing as wool and mohair. Cashgora is usually spun on the woollen system (Hunter, 2020) and is reportedly easier to spin than cashmere and mohair fibres (Hassan, 2015).

2.3.1.4. Classification of Cashgora Fibres

Cashgora fibres, after the dehairing process, are divided into three classes according to the fibre diameter as: 17-18.5, 19.5-21 and 22-23 μm (Shamey and Sawatwarakul, 2014). There are three mean types of Cashgora, ranging from the top end (18.5 μm) marketed as "Ligne Or", the medium range (20 μm) marketed as "Ligne Emerande" and the lower range (just below 22 μm) marketed as "Ligne Saphir". In Australia, the finest Cashgora (19 to 21 μm) is classified as "coarse cashmere", and the coarsest Cashgora (21 to 23 μm) is classified as "kempy mohair" (Hunter, 1993).

2.3.1.5. Microscopic Properties of Cashgora Fibres

Phan, Wortmann, and Arns (1990) examined the morphological structure of Cashgora fibres and stated that these fibres were different enough to allow them to be distinguished from cashmere fibres. In another study, Phan, Wortmann, and Arns (1991) investigated the scale characteristics of Cashgora fibres and concluded that they are closer to mohair than to cashmere but that some fibres possess cashmere-like features (cylindrical and semi-cylindrical scales) and others certain characteristics of mohair (lance-shaped scales). Vineis, Aluigi, and Tonin (2008), in their study on Cashgora found that coarse fibres showed a surface morphology very similar to that of mohair, while there were scales similar to those of cashmere on the surface of the fine fibres. In general, they stated that the cuticle cells of Cashgora have a slightly more oriented arrangement than mohair and were longer, but that there were no differences in the scale frequency and thickness, hence is would be very difficult to distinguish these fibres from each other microscopically.

The structure of the cortex of Cashgora fibres ranges from bilateral to non-bilateral, some resembling the bilateral structure of cashmere, others resembling the non-bilateral structure of mohair, with the majority being intermediate (Hunter, 1993). In a study on Australian Cashgora fibres carried out by Hudson (1992) using transmission electron microscopy (TEM), it was found that fibres with a diameter of 15.9 µm were 53.4% ortho-, 25.5% para- and 21.1% meso-cortex; while fibres with a diameter of 17.7 µm comprised 37.0% ortho-, 13.1% para- and 49.9% meso-cortex. Vineis, Aluigi, and Tonin (2008) examined the TEM microscope images of mohair (25.2 µm), cashmere (13.9 µm) and Cashgora (20.8 µm) fibres dyed with methylene blue and found that cashmere had a bilateral structure (ortho- and para-cortex), while mohair and Cashgora were homogeneously composed of ortho-cortex cells.

Although Cashgora fibres do not generally contain a medulla (Hassan, 2015), medullated fibres can be encountered in some cases (Hunter, 1993).

2.3.1.6. Physical Properties of Cashgora Fibres

Fineness (Diameter)

Cashgora fibres, which vary between 18-23 µm in diameter (Dalton and Franck, 2001), are located between cashmere and kid mohair in fineness (Dellal et al., 2010) (Figure 31). While the diameter of fine Cashgora fibres can be as low as about 12 µm, that of the coarse fibres as high as 45 µm (Hunter, 1993). The diameter variation (CV) of the fibres is around 25-30% (Hunter, 2020). New Zealand Cashgora is defined as "a subcomponent of a bilayer fleece, with an average fibre diameter of between 17.5 (sometimes as fine as 17 µm) and 23 µm, with a diameter variation of less than 28% and with fewer than 6% of the fibres being coarser than 30 µm" (Hunter, 1993).

Length

Cashgora fibres are generally 30-90 mm long (Dalton and Franck, 2001), mostly between 40 and 60 mm, and there may be fibres even exceeding 100 mm (Hunter, 2020).

Clean Down Fibre Yield

In Cashgora, the down fibres make up approximately 50% of the entire fleece (Dalton and Franck, 2001), there being less protective guard hair than in a cashmere fleece ("Animal Fibre Cashgora", 2011). Raw/greasy Cashgora contains 13.2% moisture, 1.2-2.8% grease and 0.6% water-soluble foreign matter (Dalton and Franck, 2001).

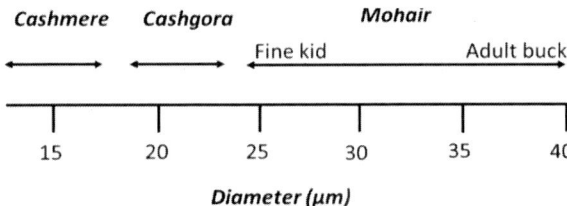

Figure 31. Diameter ranges of mohair, cashmere and Cashgora fibres (reproduced from Hunter and Hunter, 2001).

Crimp

Cashgora fibres are not crimped (Hunter, 1993), the fibre average crimp angle varying between 24 and 46°/mm, with the lowest average being 35.8°/mm. The fact that the curvature is below 45°/mm in Cashgora and greater than 45°/mm in cashmere can help to distinguish between these two fibres (McGregor, 2002).

Colour

Cashgora down fibres are usually white (Shamey and Sawatwarakul, 2014), there are, however, also light greys and browns (Ekarius and Robson, 2011).

OtherPproperties

Cashgora fibres, which are coarser and more lustrous than cashmere, feel cooler than cashmere and warmer than mohair when touched (Dellal et al., 2010). Cashgora down fibres have a low to medium lustre and generally a soft handle (Shamey and Sawatwarakul, 2014). Cashgora fibres have a significantly lower resistance to compression compared to cashmere from the new origins (McGregor, 2002).

Luxury Fibres Obtained from the Goats   117

2.3.1.7. Chemical Properties of Cashgora Fibres

Albertin, Souren, and Rouette (1990) investigated and compared the amino acid composition of New Zealand Cashgora with a diameter of 19.2 and 20.8 µm and Mongolian cashmere with a diameter of 15.9 µm. No difference was found in the amino acid compositions of the cashmere and Cashgora. They stated that the higher cysteic acid content of the finer Cashgora may result from the oxidative breakdown of the cystine constituent as a result of exposure of the fibres to ultraviolet light (UV) during growth. Table 18 gives the amino acid contents of Cashgora fibres (Tucker et al., 1988).

Table 18. The amino acid (mol%) content of Cashgora fibres (Tucker et al., 1988)

Amino acids	Australian Cashgora	
	16.2±3.3 µm	21.2±1.3 µm
Alanine	5.7	5.7
Arginine	7.4	7.8
Aspartic acid	7.1	7.1
Cysteic acid	0.1	0.2
Cystine	4.8	4.8
Glycine	8.4	7.8
Glutamic acid	13.5	13.4
Histidine	0.6	0.8
Isoloysin	3.2	3.3
Lysine	2.8	2.9
Loysin	7.7	7.5
Methionine	0.4	0.5
Proline	7.5	7.9
Serine	11.5	11.3
Phenylalanine	2.8	2.8
Threonine	6.9	7.1
Tyrosine	3,5	3,4
Valine	6.0	5.9

2.3.1.8. End-Uses of Cashgora Fibres

Cashgora fibres are more suitable for weaving than for knitting (Dalton and Franck, 2001), and are used in various clothing items (e.g., jackets, coats, scarves, shawls), apart from underwear and socks. Furthermore, blankets are also produced from these fibres (Hunter, 1993). Cashgora fibres have also been evaluated in worsted hand knitting yarns, machine knitting yarns, flame resistant clothing for motor racing and outdoor clothing (McGregor, 2012). Photos of Cashgora fibre, yarn and knitted fabric are given in Figure 32 (Ekarius and Robson, 2011).

Figure 32. Cashgora fibre, yarn and knitted fabric (adapted from Ekarius and Robson, 2011).

A few major mohair and cashmere processors believed Cashgora had good market potential and promoted their Cashgora products. William Edleston and Co. Bradford successfully marketed blends of 70% Cashgora and 30% lambswool. In Australia, Belisa Cashmere has been producing Cashgora knitwear since 1994. They reported that Cashgora was worn longer, washed more easily and pilled less than finer cashmere clothes (McGregor, 2012).

References

Agbadudu, A.B. and Ogunrin, F.O. 2006. "Aso-oke: A Nigerian classic style and fashion fabric." *Journal of Fashion Marketing and Management: An International Journal* 10(1): 97-113.

Ahmad, N. 1972. *The effect of solvents on the physical properties of mohair and viscose rayon*. PhD Thesis, Leeds University.

Akimoğlu, A. 2000. Accessed May 20. http://www.radikal.com.tr/2000/11/27/arka/01jos.shtml.

Albaugh, D. 2020. Accessed December 12. https://davethebugguy.org/2020/07/24/species-spotlight-argema-mittrei-the-madagascan-comet-moth/.

Albertin, J., Souren, I. and Rouette, H. 1990. "Cashgora or cashmere?." *Textil Praxis Inter* 45:719-723.

Alpaca. 2010. *"The Benefits of Alpaca Wool: Part 2"* Accessed December 10. https://www.alpacacollections.com/blog/benefits-alpaca-wool-part-2.html.

Alvigini, P.G. 1979. "*Le Fibre Piu Vicine Al Cielo (The Fibres Nearest to The Sky.*" First Edition, Published by A. Mondadori.

Ammayappan, L., Shakyawar, D.B., Krofa, D., Pareek, P.K. and Basu, G. 2011. "*Value addition of pashmina products: Present status and future perspectives: A review*". *Agricultural Reviews* 32(2): 91-101.

Angora Rabbit. 2020. Accessed December 10. https://en.wikipedia.org/wiki/Angora_rabbit.

Animal Fibre Cashgora. 2011. Accessed October 16. http://www.crownmountainfarms.com/html/animal-fiber/cashgora-anim.html.

Anon. 2022. *Mohair South Africa*, Port Elizabeth, South Africa.

Ansari-Renani, H.R. 2015. "Cashmere production, harvesting, marketing and processing by nomads of Iran: A review." *Pastoralism* 5(1):1-10.

Ansari-Renani, H.R., Rischkowsky, B., Mueller, J.P., Moradi, S. 2013. "Cashmere in Iran". *Animal Sciences Research Institute*, Karaj, Iran.

Antonini, M., Wang, J., Lou, Y., Tang, P., Renieri, C., Pazzaglia, I. and Valbonesi, A. 2016. "Effects of year and sampling site on mean fibre diameter of Alashan cashmere goat." *Small Ruminant Research* 137:71-72.

Appalling Abuse of Rabbits. 2013. "*Appalling abuse of rabbits on Chinese angora farms.*" Accessed October 10. https://www.galgoamigo.com/angora.html.

Arslan, Ö. 2005. "*Siirt ili köy işletmelerinde yetiştirilen tiftik keçilerinde tiftik verimleri, canlı ağırlık, vücut ölçüleri ve bu özellikler arasındaki ilişkiler* [Mohair yield, live weight, body measurements and the relations between these characteristics in Angora *goats reared in village farms in Siirt province*]." MSc Thesis., Yüzüncü Yıl Üniversitesi.

Atav, R. 2002. "*Pamuk ve yün dışındaki doğal lifler [Natural fibres other than cotton and wool*]." Graduate Thesis., Ege Üniversitesi.

Atav, R. and Demir, A. 2009. "Silk fibers gained from non-mulberry silkworms." *Electronic Journal of Textile Technologies* 3(3):56-64.

Atav, R. and Namırtı, O. 2011. "Background and the present status of the silk fibers." *SDÜ Journal of Engineering Sciences and Design* 1(3):112-119.

Atav, R. and Öktem, T. 2006. "Structural properties of mohair (Angora goat) fibres." *Tekstil ve Konfeksiyon* 16 (2):105-109.

Atav, R. unpublished. SEM photo of mohair fibre.

Atav, R., Durak, G., Öktem, T. and Seventekin, N. 2003. "Cashmere Fibers." *Tekstil ve Konfeksiyon* 13(3):115-121.

Ayaş Mohair Sock. 2008. *"Ayaş Hand Knitted Mohair Sock."* Accessed August 3 http://yunorgumodelleri.blogspot.com/2008/02/aya-tiftik-el-rgs-orap.html.

Badmaanvambuu, R., Alimaa, D., Phan, K. H., Augustin, P. and Wortmann, F. J. 2003. "Quality assessment of Mongolian cashmere: First results." *DWI Reports* (126):505-512.

Bhattacharya, T.K., Misra, S.S., Sheikh, F.D., Kumar, P. and Sharma, A. 2004. "Changthangi Goats: A rich source of pashmina production in Ladakh." *Animal Genetic Resources Information* 35:75-85.

Bohanec, H. 2017. *"Denying the freedom of flight: The story of silk"* Accessed December 13. https://www.all-creatures.org/articles/ar-denying-flight-silk.html.

Bolat, Ü. 2006. "*Adana bölgesinde yetiştirilen kıl keçilerinde alt kıl (kaşmir) üretim potansiyeli, alt kılların fiziksel karakteristikleri ve tekstil sektöründeki kullanım alanları [Down hair (cashmere) production potential of hair goats bred in Adana region, physical characteristics of down hairs and their usage areas in the textile sector*]." MSc Thesis., Çukurova Üniversitesi.

Britannica, The Editors of Encyclopaedia. Angora Goat. 2020. Accessed December 10. *Encyclopedia Britannica*, 2 May. 2013, https://www.britannica.com/animal/Angora-goat.

Burberry Cashmere. 2011. *"Burberry Cashmere Check Shawl."* Accessed August 12. http://www.eburberryoutlet.com/burberry-cashmere-check-shawl-6020-p-34.html.

Cashgora Goat. 2010. Accessed September 12. http://images.google.com.tr/imgres?imgurl =http://bp2.blogger.com/.

Cashmere Scotland. 2003. Accessed May 25. www.cashmere-scotland.co.uk/strycash.htm - 8k /.

Cashmere Women's Coat [Kaşmir Bayan Palto]. 2011. Accessed August 12. http://www.morninggloriousvintage.com/1940s1950s.html

Cashmere Wool. 2020. Accessed October 20. https://en.wikipedia.org/wiki/Cashmere_wool.

Cashmere. 2003. Accessed May 14. https://www.cashmere.com/html/101/101.html. Date of Access: 14.05.2003

References

Cashmere. 2011. Accessed October 07. *"Kaşmir Yolunda Yaşayanlar [Those Who Live on the Cashmere Road]."* http://bayankus.com/index.php/2011/02/17/kasmir-yolunda-yasananlar-1-bolum.

Chinese Oak Silkmoth. 2020. Accessed December 13. http://tpittaway.tripod.com/silk/a_per.htm.

Chiru Facts. 2011. Accessed November 05. https://www.earthislandprojects.org/tpp/chirufacts.htm.

Coloured Angora Goat. 2023. Accessed April 11 https://www.tarimtv.gov.tr/tr/video-detay/tiftik-kecisine-destekli-koruma-15394

Coloured Mohair. 2023. Accessed April 11 https://www.cagba.org/cagba-recommended-show-etiquette/breeding-for-color/

Cook, J.G. 2001. "B: Natural fibres of animal origin." In *Handbook of Textile Fibres.* Cambridge, UK: Woodhead Publishing Limited.

Corbman, B.P. 1983. *"Textiles: Fiber to fabric."* (6th edition). ISBN 0-07-013137-6, New York, ABD: McGraw-Hill, 79-165.

Cox, P. 2009. Accessed December 13. *http://artformadagascar.blogspot.com/2009/12/silk-scaves-of-ny-tanintsika.html.*

Craig, C.L., Weber, R.S. and Akai, H. 2012. "Wild silk: wild silk enterprise programs to alleviate poverty and protect habitats." In R. Kozlowski (ed.), *Handbook of Natural Fibres.* Elsevier, 576-604.

CSIRO. no date. *IWTO Publication and CSIRO/IWS Publication*, Australia.

Cricula Trifenestrata. 2020. Accessed December 13. http://silkmothsandmore.blogspot.com/2014/06/cricula-trifenestrata.html.

Currie, R. 2001. "Silk." In Robert R. Franck (ed.), *Silk, Mohair, Cashmere and Other Luxury Fibres,* 1-67, Cambridge, UK: Woodhead Publishing Limited.

Dalton, J. and Franck, R.R. 2001. *"Cashmere, camelhair and other hair fibres."* In Robert R. Franck (ed.), *Silk, Mohair, Cashmere and Other Luxury Fibres,* 133-174. Cambridge, UK: Woodhead Publishing Limited.

Davaslıgil, Ş. 1960. *"Yün ve ipek, İplik Teknolojisi [Wool and silk, Yarn Technology]."* (478-479, Volume-I), İstanbul, Türkiye: Kurtulmuş Matbaası.

Davaslıgil, Ş. 1966. *"Türkiye tiftiklerinin kıymetlendirilmesi. Kısım 1: Tiftik elyafı menşei, özellikleri, kullanıldığı yerler, işlenmesi [Evaluation of Turkish mohair. Part 1: Mohair fiber origin, properties, uses, processing]."* Ankara: TÜBİTAK Project: MAG-55.

Dellal, G., Eliçin, A., Tuncel, E., Erdoğan, Z., Taşkın, T., Cengiz, F., Ertuğrul, M., Soylemezoğlu, F., Dag, B., Ozder, M., Pehlivan, E., Tuncer Seckin, S., Kor, A., Aytac, M. and Koyuncu, M. 2010. "Türkiye'de hayvansal lif üretiminin durumu ve geleceği [Status and future of animal fibre production in Turkey]." *TMMOB Ziraat Mühendisleri Odası Türkiye Ziraat Mühendisliği VII. Teknik Kongresi,* Bildiriler Kitabı-2, 735-757, Ankara. Ocak 11-15.

Diamond Fibre Pashmina. 2009. Accessed May 14. https://www.fibre2fashion.com/industry-article/3904/diamond-fibre-pashmina.

Domestic Yak. 2020. Accessed December 10. https://en.wikipedia.org/wiki/Domestic_yak.

Ekarius, C. and Robson, D. 2011. *"The fleece & fiber sourcebook: more than 200 fibers, from animal to spun yarn."* North Adams, ABD: Storey Publishing.

Ertuğrul, M. 1991. "Küçükbaş hayvan yetiştirme uygulamaları [Small ruminant breeding practices]." *Ankara Üniversitesi Ziraat Fakültesi Yayınları* 145.

Fabric Abbreviations. 2020. Accessed September 20. https://www.joelandsonfabrics.com/uk/fabric-abbreviations.

Farming Mohair. 2020. Accessed October 20. http://www.mohairproducers.co.nz/.

Fibre Production. 2018. *"Fiber Production and Sheep Breeding in South America."* Accessed August 31. https://inta.gob.ar/sites/default/files/script-tmp inta%20fiber_production_and_sheep_breeding_in_south_ame.pdf.

Fibre Report. 2020. *"Preferred Fiber & Materials Market Report 2020"* Accessed November 22 (2021). https://textileexchange.org/app/uploads/2021/03/Textile-Exchange_Preferred-Fiber-Material-Market-Report_2020.pdf.

Fibre Report. 2021. *"Preferred Fiber & Materials Market Report 2021"* Accessed April 12 (2022). https://textileexchange.org/wp-content/uploads/2021/08/Textile-Exchange_Preferred-Fiber-and-Materials-Market-Report_2021.pdf.

Fibre Report. 2022. *"Preferred Fiber & Materials Market Report 2022"* Accessed April 15 (2023). https://textileexchange.org/app/uploads/2022/10/Textile-Exchange_PFMR_2022.pdf.

Franck, B. 2001. "Preface". In Robert R. Franck (ed.), *Silk, Mohair, Cashmere and Other Luxury Fibres.* 9-10, Cambridge, UK: Woodhead Publishing Limited.

Frank, R.H. 1999. *"Luxury fever: Weighing the cost of excess."* New Jersey, USA: Princeton University Press.

Fraser, R.D.B., MacRae, T.P. and Rogers, G.E. 1972. *"In keratins, their composition, structure and biosynthesis."* Springfield, Illinois.

Gallo, G. 2016. Accessed December 10. *https://www.magazinehorse.com/en/vicuna/*.

Gardetti, M.A. and Muthu, S.S. 2015. "The lotus flower fiber and sustainable luxury." In M.A. Gardetti and S.S. Muthu (ed.), *Handbook of Sustainable Luxury Textiles and Fashion* Section, 1, 3-18. Singapore: Springer.

Giant Silkworm. 2011. Accessed December 13. https://www.flickr.com/photos/itchydogimages/6211633662/in/photostream/.

Gingerline Image. 2011. Accessed Faburary 12. http://images.google.com.tr/imgres?imgurl=http://www.offgridlife.com/.

Goat Fibres. 2010. Accessed September 12. http://www.pcagoats.org/pcagoats_fiber_descriptions.shtml%20/.

Göktepe, F., Canipek, G. and Soysal, M.İ. 2019. "Physical characteristics of anatolian native goat down-hair fibers in comparison with cashmere fiber." *Tekstil ve Konfeksiyon* 29(1):11-21.

Göktepe, F. and Şahin, B. 2000. "Kendi anayurdunda yok olmaya yüz tutan elmas lif: Tiftik [Diamond fiber that is about to disappear in its homeland: Mohair]." *Tekstil & Teknik Dergisi* 188-192.

Grand View Research. 2019. *"Global cashmere clothing market analysis."*

Guanaco. 2020. Accessed December 10. https://en.wikipedia.org/wiki/Guanaco.

Gupta, N.P., Arora, R.K. and Patni, P.C. 1992. "Properties and processing of Angora rabbit fiber." *Indian Textile Journal* 102(7):66-72.

Günlü, A. and Alaşahan, S. 2010. "Türkiye'de keçi yetiştiriciliği ve geleceği üzerine bazı değerlendirmeler [Some evaluations on goat breeding and its future in Turkey]." *Veteriner Hekimler Derneği Dergisi* 81(2):15-20.

Gürler, A.M. 2006. "Türkiye Tiftik Cemiyeti'nin tarihçesi [History of Turkish Mohair Society]." *Lalahan Hayvancılık Araştırma Enstitüsü Dergisi* 46 (2):39-46.

Harmancıoğlu, M. 1974. *Yün ve deri ürünü diğer lifler, Lif Teknolojisi* [*Wool and other skin product fibres, Fibre Technology*], Bornova-İzmir, Türkiye: Ege Üniversitesi Ziraat Fakültesi Yayınları, (Publication no: 224)

Hassan, M. M. 2015. "Sustainable Processing of Luxury Textiles." In M.A. Gardetti and S.S. Muthu (ed.), *Handbook of Sustainable Luxury Textiles and Fashion* (Section 1, 101-120).

Hatemi, A. 1988. "*Tiftik liflerinin boyanma özelliklerinin araştırılması* [*Investigation of dyeing properties of mohair fibres*]" MSc Thesis, Ege Üniversitesi Fen Bilimleri Enstitüsü Tekstil Mühendisliği Ana Bilim Dalı, Bornova-İzmir, Türkiye.

Hayward, G. *Photo of Angora goat herd.*

Heine, K. 2012. "*The concept of luxury brands* (2nd edition)" Accessed Faberary 10. https://upmarkit.com/sites/default/files/content/20130403_Heine_The_Concept_of_ Luxury_Brands.pdf.

Heritage Cashmere. 2003. Accessed May 14. http://www.heritage-cashmere.co.uk/trade/ heritage/manu.htm/.

Hertog, J. 2020. Accessed December 13. https://tr.pinterest.com/pin/308637380698253 832/.

Hillbrick, L.K. 2012. "*Fibre Properties affecting the Softness of Wool and other Keratins.*" PhD Thesis., Deakin University, Melbourne, Victoria, Avustralya.

Historic Cashmere Markets. 2020. Accessed September 20. https://capcas.com/cashmere-history/historic-cashmere-markets/.

Horikita, M., Fukuda, M., Takaoka, A. and Kawai, H., Sen'i Gakkaishi 1989. "*Fundamental studies on the interaction between moisture and textiles, Part X: Moisture sorption properties of wool and hair fibers.*" 45:367-381.

Hudson, A.H.F. 1992. "*The chemistry and morphology of goat fibres.*" PhD Thesis., Deakin University, Geelong, Avustralya.

Huggins, D.S., Sookdeo, K. and Cock, M.J.W. 2018. "The caterpillar of Rothschildia vanschaycki (Lepidoptera, Saturniidae), a little known silk moth from Trinidad." W.I. Living World, J. *Trinidad and Tobago Field Naturalists Club*, 100-101.

Hunter, L. 1993. "*Mohair: A review of its properties, processing and applications.*" Port Elizabeth, South Africa: CSIR Division of Textile Technology.

Hunter, L. 2020. "Mohair, cashmere and other animal hair fibres." In R. Kozlowski and M. Mackiewicz-Talarczyk (ed.), *Handbook of Natural Fibres* (2nd edition, section 1, 279-383), Elsevier.

Hunter, L. and Hunter, E.L. 2001. "Mohair." In Robert R. Franck (ed.), *Silk, Mohair, Cashmere and Other Luxury Fibres* (68-117), Cambridge, UK: Woodhead Publishing Limited.

Hunter, I.M. and Kruger, P.J. 1966 "A comparison of the tensile properties of kemp, mohair and wool fibers." *SAWTRI Techn Rep*, No. 84. Port Elizabeth, South Africa.

Hunter, I.M. and Kruger, P.J. 1967. "A comparison of the tensile properties of kemp, mohair and wool fibres." *Textile Research Journal* 37:220.

Hunter, L. and Smuts, S. 1981. "Some typical bundle and single fibre tensile properties of mohair." *SAWTRI Bull*, 15(2), 18. Port Elizabeth, South Africa.

Hunter, L., Smuts, S. and Barkhuysen, F.A. 1977. "The effect of liquid ammonia treatment on some physical properties of mohair fibres." *SAWTRI Techn. Rep.*, No.372. Port Elizabeth, South Africa

Identity and Mission. 2020. Accessed September 30. https://www.cashmere.org/identity-mission.php.

İlleez, A.A., Öktem, T. and Seventekin, N. 2003. "Spider fibers." *Tekstil ve Konfeksiyon*, 13:6-10.

İmeryüz, F. 1959. "Amerika'dan gelen 6/53 tek adlı Ankara keçisi tekesinin 1,5 ve 2,5 yaşındaki yavrularıyla aynı yaşta olan Ankara Keçilerimizin beden ölçüleri, tiftik verimi, doğum ve canlı ağırlıkları üzerinde mukayeseli bir araştırma [A comparative study on body measurements, mohair yield, birth and live weights of our Ankara Goats, which are the same age as 1.5 and 2.5 years old kids of 6/53 single named Angora goat buck from America]." *Lalahan Zootekni Araştırma Enstitüsü Dergisi* 1(1):11-28.

İmeryüz, F., Müftüoğlu, Ş., Sincer, N. and Öznacar, K. 1969. "Ankara keçilerinde doğumdan itibaren ergin çağa kadar uygulanacak yılda iki kırkımın tiftik verim ve özellikleri üzerine etkisi [The effect of two shears per year on mohair yield and characteristics in Angora goats from birth to adult age]." *Lalahan Zootekni Araştırma Enstitüsü Dergisi* 9(3-4):15-33.

Kadwell, M., Fernandez, M., Stanley, H.F., Baldi, R., Wheeler, J.C., Rosadio, R., and Bruford, M.W. 2001. Genetic analysis reveals the wild ancestors of the llama and the alpaca. *Proceedings of the Royal Society Biological Sciences*. Series B. 268(1485): 2575-2584.

Karthik, T., Rathinamoorthy, R. and Ganesan, P. 2015. "Sustainable luxury natural fibers: Production, properties, and prospects." In M.A. Gardetti ve S.S. Muthu (ed.), *Handbook of Sustainable Luxury Textiles and Fashion* (Section 1, 59-98), Singapore: Springer.

Kaymakçı, M. 2006. *"Tiftik üretimi."* [Mohair production]. Keçi Yetiştiriciliği (2. bas., 9. böl.), Bornova-İzmir, Türkiye: Ege Üniversitesi.

Kaymakçı, M. and Aşkın, Y. 1997. "Keçi yetiştiriciliği [Goat farming]." İzmir, Türkiye: Baran Ofset Matbaa.

Kerven, C. and Toigonbaev, S. 2010. *"Cashmere from the Pamirs."* Accessed December 13. https://www.researchgate.net/publication/292745558_Cashmere_from_the_Pamirs.

King, N.E. 1967. "Comparison of Young's Modulus for bending and extension of single mohair and kemp fibres." *Textile research Journal* 37:204.

Krejčík, S. 2011. Accessed December 13. *https://www.biolib.cz/en/image/id161914/*.

Llama. 2020. Accessed December 10. https://en.wikipedia.org/wiki/Llama.

Leeder, J.D., McGregor, B.A. and Steadman, R.G. 1998. "Properties and performance of goat fibre." *RIRDC Publication*, No 98/22, RIRDC, Project No: ULA-8A

Leon, N.H. 1972. "Structural aspects of keratin fibres." *Journal of the Society of Cosmetic Chemists* 23:427-445.

Liu, C., Xie, C. and Liu, X. 2017. "Properties of Yak Wool in Comparison to Cashmere and Camel Hairs." *Journal of Natural Fibers* 15(2):162-173.

Liu, X. and Wang, X. 2007. "A Comparative study on the felting propensity of animal fibers." *Textile Research Journal* 77(12):957-963.

Lupton, C.J., Pfeiffer, F.A. and Blakeman, N.E. 1991. "Medullation in mohair." *Small Ruminant Research* 5(4):357-365.

Lupton, C.J. and McColl A. 2011. "Measurement of luster in Suri alpaca fiber." *Small Ruminant Research* 99:178-186.

Maasdorp, A.P.B. and Van Rensburg, N.J.J. 1983. *Proc. Conf Electron Microscopy* Soc. SA, 13, 37, South Africa

Manzano, F.J.C. 2020. Accessed December 13. *https://www.123rf.com/photo_87908 422_caterpillar-pine-processionary-species-thaumetopoea-pityocampa-on-natural-green-background.html?vti=n8af4auzhicq0m2y9n-1-2*.

Marin, J.C., Zapata, B., Gonzalez, B.A., Bonacic, C., Wheeler, J.C., Casey, C., Bruford, M.W., Palma, R.E., Poulin, E., Alliende, M.A., and Spotorno, A.E. 2007. Sistematica taxonomia y domesticacion de alpacas y llamas: nueva evidencia cromosomica y molecular (Systematics, taxonomy and domestication of alpaca and llama: new chromosomal and molecular evidence) *Revista Chilena de Historia Natural*. 80(2): 121-140.

Markova, I. 2019. *"Textile fiber microscopy: A practical approach."* John Wiley & Sons Ltd, Hoboken, ABD.

Maskiell, M. 2002. "Consuming kashmir: Shawls and empires 1500-2000." *Journal of World History* 13(1):27-65.

Mbahin, N., Raina, S.K., Kioko, E.N. and Mueke, J.M. 2010. "Use of sleeve nets to improve survival of the Boisduval silkworm, Anaphe panda, in the Kakamega Forest of western Kenya." *Journal of Insect Science* 10(6):1-10.

McGregor, B.A. 2001. *"The quality of cashmere and its influence on textile materials produced from cashmere and blends with superfine wool."* PhD Thesis., University of New South Wales, Sydney, Avustralya.

McGregor, B.A. 2002. "Australian cashmere: Attributes and processing." *A report for the Rural Industries Research and Development Corporation,* Publication No.02/112, Project No. DAV-98A, ISBN 0642 58511 3, ISSN 1440-6845.

McGregor, B.A. 2004. "Quality attributes of commercial cashmere." *South African Journal of Animal Science* 34(1):137-140.

McGregor, B.A. 2007. "Cashmere fibre crimp, crimp form and fibre curvature." *International Journal of Sheep & Wool Science* 55(1):105-129.

McGregor, B.A. 2012. *"Properties, processing and performance of rare and natural fibres: a review and interpretation of existing research results."* Rural Industries Research and Development Corporation (RIRDC), October, ISSN 1440-6845.

McGregor, B.A. 2018. "Physical, chemical, and tensile properties of cashmere, mohair, alpaca, and other rare animal fibers." In *Handbook of Properties of Textile and Technical Fibres* 105-136. Elsevier.

McGregor, B.A., Butler, K.L. and Ferguson, M.B. 2012. "The Allometric relationship between mean fibre diameter of mohair and the fleece-free live weight of Angora goats over their lifetime." *Animal Production Science* 52: 35-43.

McGregor, B.A., Butler, K.L. and Ferguson, M.B. 2013a. "The relationship between the incidence of incidence of medullated fibres in mohair and live weight over the lifetime of Angora goats." *Small Ruminant Research* 113:90-97.

McGregor, B.A., Butler, K.L. and Ferguson, M.B. 2013b. "The relationship of the incidence of medullated fibres to the dimensional properties of mohair over the lifetime of Angora goats." *Small Ruminant Research* 115:40-50.

McGregor, B.A. and Liu, X. 2017. "Cuticle and cortical cell morphology and the ellipticity of cashmere are affected by nutrition of goats." *The Journal of The Textile Institute* 108(10):1739-1746.

McGregor, B.A. and Postle, R. 2004. "Softness and other fibre attributes of commercial cashmere textiles from China and other origins of production." (372-375) In *Proceedings Textile Institute 83rd World Conferance*, Shanghai, China.

McGregor, B.A. and Quispe Peña, E.C. 2017. "Cuticle and cortical cell morphology of alpaca and other rare animal fibres." *The Journal of The Textile Institute* 109(6):767-774.

McGregor, B.A. and Schlink, A.C. 2014. "Feltability of cashmere and other rare animal fibres and the effects of nutrition and blending with wool on cashmere feltability." *The Journal of The Textile Institute* 105(9):927-937.

McGregor, B.A. and Stapleton, D.L. 2016. "Contribution of objective and subjective attributes to the variation in the whiteness and brightness of commercial mohair sale lots." *The Journal of The Textile Institute* 107(4):531-545.

McGregor, B.A. and Tucker, D.J. 2010. "Effects of nutrition and origin on the amino acid, grease, and suint composition and color of cashmere and guard hairs." *Journal of Applied Polymer Science* 117(1):409-420.

Meredith, R. 1945. "The tensile behaviour of raw cotton and other textile fibres." *Journal of The Textile Institute* 36:107-130.

Mohair Association [*Tiftik Birlik*]. 2020. Accessed June 05. http://www.tiftikbirlik.com.tr.

Mohair Beret [*Tiftik Bere*]. 2004. Accessed Faburary 14. http://www.dsk.anadolu.edu.tr/dogw/egt/mal/Malzeme.html.

Mohair Report. 2010. *"2010 yılı tiftik raporu [2010 mohair report]"*, T.C. Sanayi ve Ticaret Bakanlığı Teşkilatlandırma Genel Müdürlüğü, Accessed March 29 (2011). https://www.tgm.sanayi.gov.tr.

Mohair Report. 2013. *"2013 yılı tiftik raporu [2013 mohair report]"*, T.C. Sanayi ve Ticaret Bakanlığı Teşkilatlandırma Genel Müdürlüğü, April 20 (2014) https://www.tgm.sanayi.gov.tr.

Mohair Report. 2017. *"2017 yılı tiftik raporu [2017 mohair report]"*, T.C. Gümrük ve Ticaret Bakanlığı Kooperatifçilik Genel Müdürlüğü, Accessed April 20 (2018). https://ticaret.gov.tr/data/5d41e59913b87639ac9e02e8/2a7af69f34b40b00e7976716b7debcd3.pdf.

Mohair Report. 2018. *"2018 yılı tiftik raporu [2018 mohair report]"*, T.C. Gümrük ve Ticaret Bakanlığı Kooperatifçilik Genel Müdürlüğü, Accessed June 05 (2019).

https://ticaret.gov.tr/data/5d41e59913b87639ac9e02e8/9d8474b3c587d3aebc0a9767 3cdf2549.pdf.

Mohair Report. 2019. "*2019 yılı tiftik raporu [2019 mohair report]*", T.C. Gümrük ve Ticaret Bakanlığı Kooperatifçilik Genel Müdürlüğü, Accessed July 22 (2020). https://esnafkoop.ticaret.gov.tr/data/5d44168e13b876433065544f/2019%20Tiftik%20Raporu.pdf.pdf.

Mohair Seminar [Tiftik Semineri]. 1965, June 14-19. *Sümerbank genel müdürlüğü eğitim notları [Sümerbank general directorate training notes]*.

Mohair and Sof. 2018. "*Ankara keçisi, tiftik ve Sof [Angora goat, mohair and Sof]*.", Ankara Kalkınma Ajansı, Dumat Ofset, Ankara, Türkiye. https://www.kalkinmakutuphanesi.gov.tr/assets/upload/dosyalar/90.pdf

Mohair Story. 2004. Accessed January 26. https://www.mohairusa.com/story.html.

Mongolian Fibermark. 2020. Accessed September 30. https://trademarks.justia.com/owners/mongolian-fibermark-society-1822405.

Muga Silkworm. 2020. Accessed December 13. https://www.alamy.com/stock-photo/muga-silkworm.html.

Muskox. 2023. Accessed April 07. https://uaf.edu/lars/animals/muskox.php

Naresh, R. 2018. Accessed December 10. *http://indiasendangered.com/chinese-tourists-stopped-from-smuggling-15-shahtoosh-shawls/*.

Nosowitz, D. 2016. Accessed December 10. *https://southeastagnet.com/2016/06/08/quiviut/*.

Other Goat Families [Diğer Keçi Aileleri]. 2004. Accessed Faburary 14 http://www.ankara-tarim.gov.tr/diger/keci/akecisi.htm.

Öktem, T. and Atav, R. 2007. "Tiftik (Ankara Keçisi) liflerinin üretimi ve kullanım alanları [Production and usage areas of mohair (Ankara Goat) fibers]." *Tekstil ve Konfeksiyon* 17(1):9-14.

Öztürk, A., and Goncagül, T. 1994. "Ankara Keçilerinde doğum ağırlığı ve farklı yaşlardaki canlı ağırlığın tiftik verim ve kalitesi üzerine etkisi [The effect of birth weight and live weight at different ages on mohair yield and quality in Ankara Goats.]." *Lalahan Hayvancılık Araştırma Enstitüsü Dergisi* 34(1-2):103-109.

Pascuali Filati. 2019. Accessed December 10 (2020). https://www.pascuali.de/en/blog/knit-blog/to-know/camel-fiber.

Pashmina. 2020. Accessed August 7. https://en.wikipedia.org/wiki/Pashmina.

Phan, K.H. 2007. *Quality assessment of cashmere*, Bishkek.

Phan, K.H. and Wortmann, F.J. 2000. "Appendix 10, Quality assessment of goat hair for textile use." In: R. R. Franck (ed.), *Silk, mohair, cashmere and other luxury fiber*, Cambridge, UK: The Textile Institute, Woodhead Publishing Ltd.

Phan, K.H., Wortmann, F.J. and Arns, W. 1990. "On the morphology of Cashgora fibre." *Proc Text Conf*, DWI 105, 135, Aachen, Deutsches Wollforschungsinstitut.

Phan, K.H., Wortmann, F.J. and Arns, W. 1991. "Cashmere! Cashmere?." *Proc Text Conf*, DWI 108, 235, Aachen, Deutsches Wollforschungsinstitut.

Puchala, R., Pierzynowski, S.G., Wuliji, T., Goetsch, A.L., Sahlu, T., Lachica, M. and Soto-Navarro. S.A. 2002. "Effects of small peptides or amino acids infused to a perfused area of the skin of angora goats on mohair growth." *Journal of Animal Science* 80(4):1097-1104.

Raja, A.S.M., Shakyawar, D.B., Pareek, P.K. and Wani Sarfaraz, A. 2011. "Production and performance of pure cashmere shawl fabric using machine spun yarn by nylon dissolution process." *Indian Journal of Small Ruminants* 17(2):203-206.

Roberts, M.B. 1973. "The structure and reactivity of cashmere fibres." Phd Thesis., Department of Textile Industries, University of Leeds, Leeds, United Kingdom.

Roberts, M.B. 1977. "Some effects of common processing conditions upon mohair fibre." *SAWTRI Techn. Rep.*, No.351, Port Elizabeth, South Africa.

Samia Cynthia Drury. 2020. Accessed December 13. *https://nbair.res.in/Databases/Featured_insects/Samia-cynthia.php*.

Sea Silk. 2020. Accessed October 17. https://en.wikipedia.org/wiki/Sea_silk.

Sea Silk Weaver. 2017. *"Dünyanın tek deniz ipeği dokumacısı [The world's only sea silk weaver]"* Accessed October 17. https://www.gazetebesiktas.com.tr/2017/09/13/dunya-dunyanin-tek-deniz-ipegi-dokumacisi/.

Shahtoosh and Chirus. 2012. Accessed December 12. http://factsanddetails.com/china/cat6/sub38/item1040.html.

Shamey, R. and Sawatwarakul, W. 2014. "Innovative critical solutions in the dyeing of protein textile materials." *Textile Progress* 46(4):323-450.

Sheep and Goat. 2020. *"Türkiye'de Koyun ve Keçi Yetiştiriciliği [Sheep and Goat Breeding in Türkiye]."* Accessed October 20. http://turkiyekoyunkeci.org/tr.

Shelton, M. 1993. *"Angora goat and mohair production."* San Angelo, Texas: Anchor Publishing Company.

Shelton, M. 2007. *"Angora Goat and Mohair Production."* Texas A&M University, Texas Agricultural Experiment Station, Lecture Notes, 20-21.

Siirt Blanket. 2020. Accessed September 20. http://www.siirtbattaniyesi.net/?newUrun=1&Id=959946&CatId=bs575312&Fstate=&/%C3%87%C4%B0FT-K%C4%B0%C5%9E%C4%B0L%C4%B0K-S%C4%B0%C4%B0RT-BATTAN%C4%B0YES%C4%B0.

Silk. 2008. Accessed August 18. http://www.cs.arizona.edu/patterns/weaving/articles/sm2_silk.pdf.

Silk and Cashmere. 2002. Accessed Ocotober 05 https://www.silkandcashmere.com/news.asp.

Smith, G.A. 1988. *Rich and rare fibre differentiation. Text Technol Int*, 22.

Smuts, S., Hunter, L. and Van Rensburg, H.L.J. 1981. "Some typical single fibre tensile properties for wools produced in South Africa." *SAWTRI Techn. Rep.*, No 482, Port Elizabeth, South Africa.

Smuts, S. and Hunter, L. 1974. "A Comparison of the tenacity and extension of mohair and kemp fibres." *SAWTRI Techn. Rep.*, No. 215, Port Elizabeth, South Africa.

Susich, G. and Zagieboylo, W. 1953. "The tensile behavior of some protein fibers." *Textile Research Journal* 23(6):405-417.

Sustainable Yarn. 2016. *"Sustainable yarn is a global trend: Hungaro-Len features in WGSN report"* Accessed October 10. https://www.hungarolen.hu/en/2016/05/25/sustainable-yarn-is-a-global-trend-hungaro-len-features-in-wgsn-report/.

Süpüren Mengüç, G. and Özdil, N. 2014. "Speciality animal fibers." *Electronic Journal of Textile Technologies* 8(2):30-47.

References

Şahin, G. 2013a. "Türkiye'de Ankara Keçisi (*Capra hircus Ancryrensis*) yetiştiriciliğinin dünü bugünü ve yarını [The past, present and future of Angora Goat (*Capra hircus Ancryrensis*) breeding in Türkiye]" *Manisa Celal Bayar Üniversitesi Sosyal Bilimler Dergisi* 11(2):338-352.

Şahin, G. 2013b. "Coğrafi bir simge olarak Ankara Keçisinin Türkiye'deki mevcut durumu [The current situation of the Angora Goat in Turkey as a geographical symbol]." *Milli Folklor* 25(97):195-209.

Şengonca, M. and Koşum, N. 2005. "*Koyun ve keçi yetiştirme (keçi yetiştirme ve ıslahı).*" [*Sheep and goat breeding (goat breeding and reclamation)*] Ege Üniversitesi Ziraat Fakültesi Zootekni Bölümü, İzmir.

Tarakçıoğlu, I. 1983. Tekstil terbiyesi ve makinaları, Cilt II *Protein (Yumurta akı) liflerinin terbiyesi* [Textile finishing and machinery, Volume II Finishing of protein (egg white) fibres], Bursa: Uludağ Üniversitesi Basımevi.

Tester, D.H. 1987. "Fine structure of cashmere and superfine Merino wool fibres." *Textile Research Journal* 57:213-219.

The Angora Goat. 2011. "The die Angora (Autumn 2011)" *The Angora Goat & Mohair Journal* 53(1):1-81.

The Schneider Group. 2020. Accessed September 20. https://www.gschneider.com/.

Tortora, R.G. and Johnson, I. 2013. "*The fairchild books dictionary of textiles.*" (8th edition), New York: Bloomsbury Publishing.

Tucker, D.J., Hudson, A.H.F., Ozolins, G.V., Rivett, D.E. and Jones, L.N. 1988. "Some Aspects of the Structure and Composition of Speciality Animal Fibres." *Proc 2nd Symp Speciality Animal Fibres*, Aachen, DWI 103, 71.

Tucker, D.J., Hudson, A.H.F., Rivett, D.E. and Logan, R.I. 1990. "The chemistry of speciality animal fibres." *Proc. 2nd Int. Symp. Speciality Animal Fibers*, Aachen, Deutsches Wollforschungsinstitut, DWI 106, 1.

Tunçel, K.Ş, Taşkaynatan, H. and Erkuzu, D. 2020. "Siirt battaniyesi ve teknik özellikleri [Siirt blanket and its technical features]." *Akademik Sanat Tasarım ve Bilim Dergisi* 5(9):47-60.

Utkanlar, N., İmeryüz, F., Müftüoğlu, S. and Öznacar, K. 1964. "Ankara keçilerinde yılda iki kırkımın tiftik verimi, kalitesi ve yavru verimi üzerine etkileri [The effects of two shears per year on mohair yield, quality and kidding yield in Angora goats]." *Lalahan Zootekni Araştırma Enstitüsü Dergisi* 4(4): 200-212.

Van der Westhuysen, J.M. 2005. "Marketing goat fibres." *Small Ruminant Research* 60:1-2, 215–218.

Van der Westhuysen, J.M., Wentzel, D. and Grobler, M.C. 1988. "*Angora Goats and Mohair in South Africa.*" Third edition, NMB Printers Port Elizabeth, South Africa.

Van Rensburg, N.J.J. and Maasdorp, A.P.B. 1985. "A study of the lustre of mohair fibres." *International Wool Textile Research Conference,* Tokyo, Vol. 1, 243-252.

Van Rensburg, N.J.J. 1978. "A note on the light degradation of mohair." *SAWTRI Bull.* 12(4), 48, Port Elizabeth, South Africa.

Vatansever, H. 2004. "*Lalahan Hayvancılık Merkez Araştırma Enstitüsü'nde yetiştirilen farklı kökenli Ankara keçilerinde büyüme, döl verimi ve tiftik özellikleri* [Growth, fertility and mohair characteristics of Angora goats of different origins reared in

Lalahan Animal Husbandry Central Research Institute]." PhD Thesis., Ankara Üniversitesi, Sağlık Bilimleri Enstitüsü, Ankara, Türkiye.

Vicuña. 2020. Accessed October 12. https://en.wikipedia.org/wiki/Vicu%C3%B1a.

Vineis, C., Aluigi, A. and Tonin, C. 2008. "Morphology and thermal behaviour of textile fibres from the hair of domestic and wild goat species." *AUTEX Research Journal* 8(3):68-71.

Von Bergen, W. and Krauss, W. 1942. "A collection of photomicrographs of common textile fibers." *Textile Fiber Atlas*, New York, ABD: American Wool Handbook Company.

Wallace, J. 2008. Date of Accessed December 10 (2020). *https://www.grit.com/animals/get-your-goat?SlideShow=1*.

Wang, L., Singh, A. and Wang, X. 2008. A study on dehairing australian greasy cashmere. *Fibers and Polymers* 9(4):509-514.

Ward, W.H., Binkley, C.H. and Snell, N.S. 1955. "Amino acid composition of normal wools, wool fractions, mohair, feather, and feather fractions." *Textile Research Journal* 25(4):314-325.

Wardeh, M.F., and Dawa, M. 2004. "Camels and Dromedaries: General Perspectives" In R. Cardellino, A. Rosati & C. Mosconi (ed.) *Current Status of Genetic Resources, Recording and Production Systems in African, Asian and American Camelids, FAO-ICAR Seminars on Camelids* (Sousse, Tunisia, 30 May), ICAR Technical Series, No. 11, 1-9

Weijer, F. 2011. Accessed August 07. *Cashmere Value Chain Analysis Afghanistan*, http://www.ahdp.net/reports/Cashmere%20Value%20Chain%20Analysis.pdf.

Wildman, A.B. 1954. "The Microscopy of Animal Textile Fibres (Chapter 10)", *Wool Ind. Res. Assoc*, Leeds, 106.

Wild Silk. 2020. Accessed December 13. https://www.alamy.com/stock-photo/wild-silk.html.

Women's Cashmere. 2011. Accessed August 10 https://www.vivi-direct.com/collections/womens-cashmere.

Wool [*Yapağı*]. 2011. Accessed July 04. https://zootekni.comu.edu.tr/class/hub/04%20Yapagi.pdf.

World Bank. 2019. *Mongolia Central Economic Corridor Assessment, A Value Chain Analysis of the Cashmere-Wool, Meat, and Leather Industries.* © World Bank. Washington, ABD: The World Bank Group. Accessed Jaunary 20. http://documents1.worldbank.org/curated/en/951491558704462665/pdf/Mongolia-Central-Economic-Corridor-Assessment-A-Value-Chain-Analysis-of-Wool-Cashmere-Meat-and-Leather-Industries.pdf.

Yang, G., Fu, Y., Hong, X. and Wang, C. 2005. "Discussion on cashmere fiber identification technique by SEM and LM." In *Proceedings of the 3rd International Cashmere Determination Technique Symposium* (189-205). China National Cashmere Products Engineering and Technical Centre and China Inner Mongolia Erdos Cashmere Group Corporation, Erodosi, China.

Yazıcıoğlu, G. 1996. "*Tekstil mikroskobisi* [*Textile microscopy*]." İzmir, Türkiye: Ege Üniversitesi Basımevi.

Yazıcıoğlu, G. and Gülümser, G. 1993. *"Tekstil lifleri [Textile Fibres]."* Bursa: Etin Kitapevi.

Yertürk, M. 1998. *"Doğu ve Güneydoğu Anadolu bölgesinde yetiştirilen renkli tiftik keçilerinin yarı entansif şartlarda verim özelliklerinin araştırılması [Investigation of yield characteristics of pigmented Angora goats reared in Eastern and Southeastern Anatolia under semi-intensive conditions]."* PhD Thesis, Yüzüncü Yıl Üniversitesi, Van, Türkiye.

Yondonsambuu, G., and Altantsetseg, D. 2003. *"Survey on production and manufacturing of the wool, cashmere, and camel hair. Mongolian Wool and Cashmere Association"*, Accessed Faburary 18. https://www.yumpu.com/en/document/read/11551989/survey-on-production-and-manufacturing-of-the-wool-cashmere-and-.

Yurdakul, A., and Atav, R. 2006. *"Boya baskı esasları [Fundamentals of dyeing and printing]."* Bornava, İzmir: Ege Üniversitesi Mühendislik Fakültesi Tekstil Mühendisliği Bölümü.

Ziegler, H. 2007. Accessed December 13. *http://www.lepiforum.de/2_forum.pl?md=read; id=16985.*

Index

A

accessories, 70, 75
amino acid, 15, 48, 66, 105, 106, 107, 117, 126, 127, 130
angora goat, xv, 1, 4, 8, 19, 20, 21, 22, 23, 24, 25, 30, 31, 32, 33, 34, 35, 36, 37, 39, 43, 57, 62, 63, 65, 66, 75, 77, 79, 84, 111, 113, 120, 123, 124, 126, 127, 128, 129, 131

B

bending, 56, 99, 124
blankets, 69, 70, 71, 76, 117

C

carpets, 59, 69, 70, 71, 77, 110
Cashgora, 5, 7, 19, 92, 111, 112, 113, 114, 115, 116, 117, 118, 119, 120, 127
cashmere, xiii, xiv, xvi, 1, 2, 3, 4, 7, 8, 10, 19, 20, 35, 77, 78, 79, 80, 81, 82, 83, 84, 85, 86, 87, 88, 89, 90, 91, 92, 93, 94, 95, 96, 97, 98, 99, 100, 101, 102, 103, 104, 105, 106, 107, 108, 109, 110, 111, 112, 113, 114, 115, 116, 117, 118, 119, 120, 121, 122, 123, 124, 125, 126, 127, 128, 129, 130, 131
cashmere goat, 1, 2, 4, 19, 20, 35, 77, 78, 79, 80, 82, 83, 84, 85, 87, 88, 90, 91, 93, 96, 97, 98, 99, 104, 105, 110, 111, 113, 119
Çengelli, 62
chemical(s), xvi, 2, 9, 11, 48, 56, 65, 66, 105, 106, 117, 125, 138
classification, xvi, 37, 39, 41, 42, 43, 58, 63, 92, 93, 114
clean fibre yield, 49, 96
clothing, xiv, xvi, 3, 5, 17, 32, 34, 68, 69, 73, 74, 80, 107, 108, 110, 117, 122
coarse guard hair, 10, 19, 77, 78, 87, 90, 91, 93, 94, 96, 97, 100, 105, 110, 113, 114
colour, 3, 8, 15, 22, 31, 33, 34, 40, 49, 57, 58, 59, 60, 62, 63, 76, 79, 83, 84, 89, 92, 93, 94, 96, 101, 102, 103, 106, 107, 116, 138
combing, 2, 9, 10, 47, 70, 73, 81, 82, 83, 87, 88, 89, 91, 100
cortex, 44, 45, 46, 62, 66, 67, 93, 94, 115
crimp, 11, 61, 62, 67, 89, 101, 104, 105, 116, 125
cross-sections, 50, 94
cuticle, 44, 45, 46, 57, 93, 95, 99, 107, 115, 126

D

dehairing, xvi, 10, 89, 90, 91, 92, 96, 97, 99, 100, 109, 114, 130
diameter, 7, 9, 10, 37, 38, 41, 46, 47, 49, 50, 51, 52, 55, 60, 62, 65, 67, 78, 90, 92, 93, 94, 95, 96, 97, 98, 99, 100, 101, 102, 103, 104, 105, 112, 114, 115, 116, 117, 119, 126

E

elongation (elasticity), 49
end-uses, xvi, 48, 68, 69, 72, 107, 117
epidermis cells, 95

Index

eumelanin, 62, 63, 64

F

fabric(s), xvi, 1, 8, 10, 17, 19, 20, 23, 24, 32, 49, 54, 59, 60, 65, 68, 69, 71, 73, 74, 75, 76, 78, 80, 89, 108, 109, 110, 117, 118, 119, 121, 122, 128, 137, 138
felting, 22, 35, 45, 57, 60, 65, 104, 125
fine down fibres, 8, 10, 19, 77, 82, 87, 90, 92, 93, 94, 96, 97, 105, 110, 113, 114
fineness (diameter) and its variation, 49
fleece, 20, 23, 37, 38, 52, 57, 58, 59, 61, 87, 88, 89, 90, 99, 102, 113, 114, 115, 116, 121, 126
follicles, 8, 19, 21, 35, 36, 47, 50, 87, 88, 113

G

gingerline, 43, 62, 63, 122
goat, xvi, 2, 5, 6, 19, 20, 21, 22, 23, 24, 25, 26, 31, 32, 33, 34, 35, 36, 37, 39, 41, 48, 50, 51, 52, 53, 55, 56, 61, 63, 68, 70, 72, 74, 78, 80, 82, 84, 85, 87, 88, 91, 92, 97, 100, 110, 111, 112, 113, 120, 121, 122, 123, 124, 127, 128, 129, 130
guard hair, 19, 77, 87, 89, 90, 91, 92, 94, 98, 99, 100, 105, 107, 110, 113, 114, 116, 126

H

harvesting, xvi, 2, 3, 9, 35, 83, 87, 89, 100, 113, 119
hybrid(s), 19, 84, 111, 112

I

inorganic, 57, 59
interlinings, 71, 75

K

kemp, 33, 34, 37, 38, 40, 41, 45, 46, 47, 48, 49, 50, 56, 59, 60, 64, 77, 94, 123, 124, 128

keratin, xv, 9, 11, 15, 46, 48, 65, 67, 105, 107, 125
knitwear, 72, 82, 102, 108, 109, 110, 118

L

layer(s), 8, 10, 43, 44, 45, 46, 47, 48, 57, 62, 64, 77, 89, 94, 113, 138
length and its variation, 49
lining(s), 69, 75
lustre, 3, 11, 12, 22, 40, 45, 49, 57, 60, 65, 67, 70, 74, 116, 129

M

medulla, 10, 45, 46, 47, 48, 50, 56, 59, 93, 115
medullation/kemp, 49
menswear, 69, 71, 73
microscopic, xvi, 10, 43, 93, 114
mohair, xiii, xv, xvi, 1, 3, 7, 8, 9, 10, 11, 19, 20, 21, 22, 23, 24, 25, 26, 27, 28, 30, 31, 32, 33, 34, 35, 36, 37, 38, 39, 40, 41, 42, 43, 44, 45, 46, 47, 48, 49, 50, 51, 52, 53, 54, 55, 56, 57, 58, 59, 60, 61, 62, 63, 64, 65, 66, 67, 68, 69, 70, 71, 72, 73, 74, 75, 76, 77, 85, 94, 95, 103, 111, 112, 113, 114, 115, 116, 118, 119, 120, 121, 122, 123, 124, 125, 126, 127, 128, 129, 130, 139
mohair Mark, 70, 71
moisture, 22, 57, 58, 64, 68, 69, 72, 100, 106, 116, 123

O

ondulation (waviness), 49
origin, xv, 4, 5, 7, 12, 14, 19, 22, 62, 79, 80, 82, 99, 104, 105, 106, 108, 121, 126

P

pashmina, 78, 97, 119, 120, 121, 127
pheomelanin, 62, 63, 64
pigment granules, 94
production, xiii, xvi, 1, 2, 3, 5, 7, 8, 11, 12, 14, 15, 17, 20, 24, 25, 26, 27, 28, 30, 32,

33, 34, 36, 50, 52, 54, 68, 69, 72, 74, 75, 76, 78, 81, 82, 83, 84, 85, 87, 90, 91, 93, 95, 102, 104, 107, 109, 110, 113, 114, 119, 120, 121, 122, 124, 126, 127,128, 130, 131

S

scale frequency, 10, 95, 104, 115
scouring, 11, 33, 58, 61, 65, 67, 83, 89, 91, 99, 100, 107
shawls, scarves, 69, 108, 109
shearing, 2, 9, 20, 33, 35, 36, 37, 38, 39, 40, 42, 52, 53, 54, 57, 59, 65, 68, 87, 88, 89, 100
softness, xiv, 1, 3, 10, 19, 57, 65, 78, 80, 95, 97, 103, 104, 105, 108, 123, 126
sorting, 38, 89, 91
spring (*Bahari*) cashmere, 93
strength, 12, 15, 22, 49, 51, 54, 55, 56, 67, 68, 98, 107, 108
strength and its variation, 49

T

tannery (or skin) cashmere, 93

U

undercoat, 3, 10, 19, 77, 78, 87, 98, 114
upholstery, 68, 69, 70, 71, 76

V

vegetable, 17, 38, 42, 49, 57, 58, 59, 88, 89, 92, 100
vegetable and inorganic matter (such as dirt and stain) content, 49

W

waviness, 11, 61
willowing, 89, 91
wool, xiv, xv, 1, 2, 3, 5, 7, 8, 9, 10, 11, 15, 16, 17, 20, 23, 30, 36, 38, 45, 46, 47, 50, 53, 55, 56, 57, 58, 60, 61, 64, 65, 66, 67, 68, 69, 70, 72, 73, 74, 75, 76, 77, 78, 80, 82, 85, 90, 92, 93, 95, 96, 98, 99, 101, 103, 104, 105, 106, 107, 108, 109, 112, 114, 119, 120, 121, 123, 124, 125, 126, 129, 130, 131, 138, 139

About the Authors

Dr. Rıza Atav, Prof.

Rıza ATAV was born on 21.09.1981 in İzmir-Türkiye. He graduated from İzmir Private Turkish High School in 1998 as a highest ranked student. He started his university education in the same year and graduated from Ege University Faculty of Engineering, Textile Engineering Department with the title of Textile Engineer as the top student of the department in 2002. In the same year, he started to work as a research assistant and began his postgraduate education at Ege University. He received MSc degree in 2005 and PhD degree in 2009. Afterwards, he started to work as assistant professor at Tekirdağ Namık Kemal University in August of the same year. He was awarded the title of associate professor in 2012. Since December 2017, he has been a full professor at Tekirdağ Namık Kemal University Textile Engineering Department.

He speaks fluent English, good Italian, intermediate Spanish and novice level French. In 2008, he won the Italian Language Education Scholarship at the Universita˙ per stranieri di Siena, ranking first in the exam held by the İzmir Italian Cultural Centre.

His main research and interest fields are textile dyeing-printing, functional-intelligent textiles and luxury animal fibres. For more than 10 years, he has been giving undergraduate courses namely Dyeing Technology I, Dyeing Technology II, Textile Printing, Textile Chemistry, Finishing of Denim Fabrics, Professional English, and Italian for Engineers; master courses namely Functional Textiles, Luxury Hair Fibres and Their Finishing Processes, Luxury Secretion Fibres and Their Finishing Processes, and New Technologies and Methods in Textile Dyeing; and PhD courses namely Scientific Basis of Textile Dyeing, and Applications of Enzymes and Nanotechnological Products in Textile Dyeing.

He has more than 100 articles published in various national and international journals, more than 100 papers presented at national and international congresses, 2 national books, 4 international book chapters. In addition, he has more than 50 completed and ongoing projects in total. He carries out tasks such as referee, scientific advisory board member, editorial board member in more than 30 national and international journals. He has also been the guest editor for the special issue of "Angora Fibre" published by "American Journal of Materials Engineering and Technology" in 2014 and special issue of "3rd International Congress of Innovative Textiles & 2nd International Congress on Wool and Luxury Fibres" published by "Coloration Technology" in 2023. He has graduated 14 MSc students so far and he is currently consulting 5 MSc students and 3 PhD students. His scientific studies were cited over 750 times in various journals and books.

He carried out various administrative duties such as Vice-Head of Department, Vice Dean, Head of Department, and various commission memberships. He took part in the congress organization committee of the International Innovative Textiles Congress (ICONTEX) held in 2011, 2019 and 2022. In 2019 and 2022, he held the International Wool and Luxury Fibres Congress (ICONWOOLF) that over 100 people attended consisting of academicians and students from various countries and representatives of various Turkish public institutions and textile industry enterprises.

He is currently working at Tekirdağ Namık Kemal University Textile Engineering Department as a full professor and is head of the department. At the same time, he carries out consultancy activities in the R&D centre of various companies within the scope of university-industry cooperation. His R&D studies in the field of wool and luxury fibres are as follows; but not limited to:

- examining various properties of these fibres
- examining various properties of fabrics produced from these fibres,
- washing and bleaching processes of these fibres,
- adding functional properties to fabrics made from these fibres,
- saving water, time, energy, dye and chemicals in dyeing these fibres,
- adapting new dyeing methods (layer-by-layer etc.) to these fibres,
- using new technologies (ozone, ultrasound, plasma, etc.), enzymatic treatments and biotechnological products (dendrimer, liposome, etc.) in dyeing these fibres,
- obtaining various colour effects in dyeing these fibres,
- obtaining colour in these fibres without using dyes,
- reuse of the dyeing wastewater of these fibres and
- decolourisation of dyeing wastewater of these fibres

His academic publications on wool and luxury animal fibres can be reached from his website.

Dr. Lawrance Hunter, Prof.

Prof. Lawrance Hunter, PhD, C Text FTI, Honourary Professor and Head of the Department of Textile Science at the Nelson Mandela University (Gqeberha/Port Elizabeth, South Africa) has been involved in textile R&D and academia for some 57 years during which time he has undertaken pioneering research and also published widely, notably in the fields of wool and mohair.

He was admitted at the Fellow of the Textile Institute (UK) in 1979 and in 1994. He received its prestigious Warner Memorial Medal in recognition of outstanding published work in textile science and technology.